Human Heredity

International Journal of Human and Medical Genetics

Founded 1950 as 'Acta Genetica et Statistica Medica' by Gunnar Dahlberg
Continued by M. Hauge (1965–1983)

KARGER

Printed in Switzerland
on acid-free paper by
Reinhardt Druck, Basel

Appears 6-weekly:
2 volumes per year
(8 issues)

Guidelines for Authors

Human Heredity

Invitation for Papers

Only original papers written in English will be considered.

Electronic submission preferred.
For rapid turnaround, manuscripts may be submitted by e-mail to HH@linkage.rockefeller.edu. E-mail submission should come as files attached to e-mail, in a format that can be read by MS Word 97 (preferred), or as a PDF or postscript. If LaTeX is used as the word processing program, please send original LaTeX data format.

Hard copy submission
Manuscripts can be sent to:

Dr. Derek Gordon
Managing Editor
The Rockefeller University
1230 York Avenue, Box 192
New York, NY 10021 (USA)

Manuscripts should be submitted in triplicate (with 3 sets of illustrations of which 1 is an original), typewritten double-spaced on one side of the paper, with a wide margin. The final version of your manuscript and illustrations should also be submitted on disk. For details refer to 'Disk Submission' or www.karger.com/journals/disksub.htm. Editorial review and copyediting will be executed from the disk version of your manuscript.

Conditions

'Human Heredity' publishes papers reporting on original investigations in the field of human and medical genetics. The journal will consider original papers, short communications and mutation reports as well as articles for the section 'Methodological Issues'. This section is intended for reports of a more technical nature. They need not in principle represent original research but may explain unusually difficult or little known problems, for example in segregation or linkage analysis. All manuscripts are subject to editorial review. Manuscripts are received with the explicit understanding that they are not under simultaneous consideration by any other publication. Submission of an article for publication implies the transfer of the copyright from the author to the publisher upon acceptance. Accepted papers become the permanent property of 'Human Heredity' and may not be reproduced by any means, in whole or in part, without the written consent of the publisher. It is the author's responsibility to obtain permission to reproduce illustrations, tables, etc. from other publications.

Short communications: Communications not exceeding 2 printed pages will usually appear in print within 3–5 months.

Mutation reports: Mutation reports should not exceed 2 printed pages and are arranged as follows: Title (including name of mutation and gene); Authors; Affiliations; Description of the mutation; Source of material (clinical and ethnic data); Methods used; Evidence for disease association; Segregation data; Frequency (among disease alleles or in population); Other Comments; Acknowledgements; References (3 at most). Illustrations are welcome if fitting within the 2 printed pages.

Arrangement

Title page: The first page of each paper should indicate the title, the authors' names, the institute where the work was conducted, and a short title for use as running head.

Full address: The exact postal address of the corresponding author complete with postal code must be given at the bottom of the title page. Please also supply phone and fax numbers, as well as e-mail address.

Key words: For indexing purposes, a list of 3–10 key words in English is essential.

Abstract: **Each paper needs an abstract of up to 200 words structured with subheadings as follows: Objective(s), Methods, Results, Conclusion(s).**

Nomenclature, abbreviations and symbols: Authors should use Human Gene Mapping (HGM)-approved genetic notation and symbols for human genes. See Shows et al: Guidelines for human gene nomenclature. Cytogenet Cell Genet 1987;46:11–28 / McAlpine et al: Report of the nomenclature committee and the 1989 catalog of mapped genes. Cytogenet Cell Genet 1989;51:13–66 / Kidd et al: Report of the DNA committee and catalogs of cloned and mapped genes and DNA polymorphisms. Cytogenet Cell Genet 1989;51: 622–947. Symbols for human genes not included in the above may be obtained from P.J. McAlpine (Co-Chair HGM Nomenclature Committee), Department of Human Genetics, University of Manitoba, T250-770 Bannatyne Avenue, Winnipeg, Manitoba, Canada R3E OW3; Telephone 204/788-6393, Fax 204/ 786-8712, BITNET: GENMAP@ UOFMCC. Authors should use SI units and follow the guidelines for abbreviation and symbols of the IUPAC-IUB Joint Commission on Biochemical Nomenclature. See Eur J Biochem 1970;15:203–208, corrected 1972;25:1, and for bases in nucleic acid sequences, Eur J Biochem 1985;150:1–5. All nonstandard abbreviations used in the text should be defined immediately before the first such abbreviation. For the mouse, see Lyon MF, Searle AG: Genetic Variants and Strains of the Laboratory Mouse. London, Oxford University Press, 1989. Approved human gene symbols should be obtained prior to submission from Dr. Sue Povey, HUGO Gene Nomenclature Committee (HGNC), University College London, Wolfson House, 4 Stephenson Way, London, NW1 2HE, UK; Tel. +44 20 7679 7410; Fax +44 20 7387 3496; E-Mail: nome@galton.ucl.ac.uk; URL: http://www.gene.ucl.ac.uk/nomenclature/ Approved mouse nomenclature should be obtained prior to submission from Lois Maltais, The Jackson Laboratory, Bar Harbor, ME 04609-0800, USA; Tel. +1 207 288 6429; Fax +1 207 288 6132; E-Mail: nomen@informatics.jax.org; URL: http://www.informatics.jax.org/mgihome/nomen/.

Statistics: In giving the frequency of polymorphisms, e.g. RFLPs, authors should state the number of chromosomes assayed in unrelated persons and the ethnic background of the subjects.

Footnotes: Avoid footnotes. When essential, they are numbered consecutively and typed at the foot of the appropriate page.

Tables and illustrations: Tables and illustrations (both numbered in Arabic numerals) should be prepared on separate sheets. Tables require a heading and figures a legend, also prepared on a separate sheet. For the reproduction of illustrations, only good drawings and original photographs can be accepted; negatives or photocopies cannot be used. Due to technical reasons, figures with a screen background should not be submitted. When possible, group several illustrations on one block for reproduction (max. size 181 × 223 mm) or provide crop marks. On the back of each illustration, indicate its number, the author's name, and 'top' with a soft pencil. Electronically submitted b/w half-tone and color illustrations must have a final resolution of 300 dpi after scaling, line drawings one of 800–1200 dpi.

Color illustrations: Up to 6 color illustrations per page can be integrated within the text at the special price of CHF 660.– / USD 545.00 per page. Color illustrations are reproduced at the author's expense.

References: In the text identify references by Arabic numerals [in square brackets]. Material submitted for publication but not yet accepted should be noted as 'unpublished data' and not be included in the reference list. The list of references should include only those publications which are cited in the text. Do not alphabetize; number references in the order in which they are first mentioned in the text. The surnames of all authors followed by initials should be given. There should be no punctuation other than a comma to separate the authors. Preferably, please cite all authors. Abbreviate journal names according to the Index Medicus system. (Also see International Committee of Medical Journal Editors: Uniform requirements for manuscripts submitted to biomedical journals. N Engl J Med 1997;336:309–315.)
This is the preferred style of references. Other styles may be applied, but should be used consistently.

Digital Object Identifier (DOI)
S. Karger Publishers supports DOIs as unique identifiers for articles. A DOI number will be printed on the title page of each article. DOIs can be useful in the future for identifying and citing articles published online without volume or issue information. More information can be found at www.doi.org.

Examples
(a) Papers published in periodicals: Sun J, Koto H, Chung KF: Interaction of ozone and allergen challenges on bronchial responsiveness and inflammation in sensitised guinea pigs. Int Arch Allergy Immunol 1997;112:191–195.
(b) Papers published only with DOI numbers:
Theoharides TC, Boucher W, Spear K: Serum interleukin-6 reflects disease severity and osteoporosis in mastocytosis patients. Int Arch Allergy Immunol DOI: 10.1159/000063858.
(c) Monographs: Matthews DE, Farewell VT: Using and Understanding Medical Statistics, ed 3, revised. Basel, Karger, 1996.
(d) Edited books: Parren PWHI, Burton DR: Antibodies against HIV-1 from phage display libraries: Mapping of an immune response and progress towards antiviral immunotherapy; in Capra JD (ed): Antibody Engineering. Chem Immunol. Basel, Karger, 1997, vol 65, pp 18–56.

Page charges

There is a page charge of CHF 55.–/USD 35.00 per page (including tables, illustrations and references). One printed page is equal to approximately 3 manuscript pages.

Galley proofs

Unless indicated otherwise, galley proofs are sent to the first-named author and should be returned with the least possible delay. Alterations made in galley other than the correction of printer's errors, are charged to the author. No page proofs are supplied.

Reprints

Order forms and a price list are sent with the galley proofs. Orders submitted after the issue is printed are subject to considerably higher prices.

KARGER

Fax +41 61 306 12 34
E-Mail karger@karger.ch
www.karger.com

The Guidelines for Authors are available at:
www.karger.com/hhe_Guidelines

Human Heredity

General Information

Publication data: 'Human Heredity' is published 8 times a year. Volumes 55–56, each with 4 issues, appear in 2003.

Subscription rates: Subscriptions run for a full calendar year. Prices are given per year.

Personal subscription:

Print or Online	Print/Online combined
CHF 678.–	CHF 756.–
EUR 484.–	EUR 540.–
USD 521.00	USD 581.00

postage and handling (added to print and print/online)
CHF 40.40 Europe, CHF 60.– Overseas
EUR 29.–
USD 40.40

Institutional subscription:

Print or Online	Print/Online combined
CHF 1356.–	CHF 1464.–
EUR 968.–	EUR 1046.–
USD 1042.00	USD 1126.00

postage and handling (added to print and print/online)
CHF 50.40 Europe, CHF 74.80 Overseas
EUR 36.–
USD 50.40

Airmail surcharge: CHF 50.40 / USD 42.00

Microforms: Please direct all enquiries and orders to:
University Microfilms Inc.
300 North Zeeb Road, P.O. Box 1346
Ann Arbor, MI 48106-1346 (USA)

Subscription orders: Orders can be placed at agencies, bookstores, directly with the
Publisher S. Karger AG, P.O. Box,
CH–4009 Basel (Switzerland)
or with any of the following distributors:
Deutschland: S. Karger GmbH, Postfach
D–79095 Freiburg
France: Librairie Luginbühl
36, bd de Latour-Maubourg
F–75007 Paris
Great Britain, Ireland: S. Karger AG
4 Rickett Street
London SW6 1RU, UK
USA: S. Karger Publishers, Inc.
26 West Avon Road, P.O. Box 529
Farmington, CT 06085, USA
India, Bangladesh, Sri Lanka:
Panther Publishers Private Ltd.
33, First Main
Koramangala First Block
Bangalore 560 034, India
Japan: Karger Japan, Inc.
Yushima S Bld. 3F
4-2-3, Yushima, Bunkyo-ku
Tokyo 113-0034, Japan

Change of address: Both old and new address should be sent to the subscription source.

Single issues: Information can be obtained through the Publisher.

Advertising: Correspondence and rate requests should be addressed to the Publisher.

Bibliographic indices: This journal is regularly listed in bibliographic services, including *Current Contents®*.

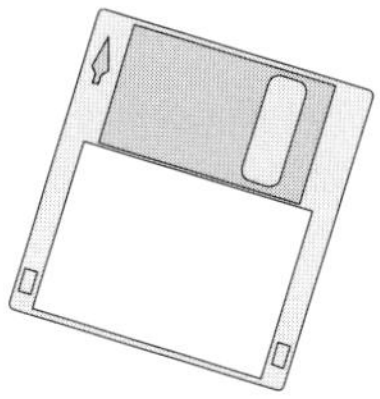

Disk Submission

A **comprehensive version** of these guidelines for the **submission of manuscripts and illustrations on disk** can be accessed online under **www.karger.com/disksubmission** or requested as a leaflet from the editorial department.

Please follow the general rules on style and arrangement in the **'Guidelines for Authors'** of the journal. Submit a **double-spaced printout** of your manuscript matching the file on disk exactly.

Technical Information

Store only data pertinent to this manuscript. Include your fonts (screen and printer fonts for Macintosh). For a **text only** manuscript, use a 3.5-inch disk. For **text and illustrations,** use 3.5 or 5.25 Magneto Optical Disc; CD-ROM (DOS or hybrid); iomega Zip/DOS (100 MB), or 44 MB, 88 MB or 200 MB SyQuest. Compress extensive data using ZIP format.

Data Description

Label your disk with the **title of the journal or book** for which it is submitted; a **short version** of the paper's title; your **name,** postal and e-mail address, fax and phone number; **hard- and software** used, including the version; and the **complete list of the file names.** File names consist of eight characters before the dot identifying the file and three characters after the dot identifying the application.

Text and Format

The **preferred word-processing package is MS-Word.** Other commonly used PC text programs and ASCII formats are accepted.
Update your data when you revise subsequent to the reviewer's or editor's decision and mark revisions clearly. Manuscripts accepted for publication are subject to **copy editing:** changes and/or questions will be marked in the manuscript, which will be returned to you with your disk and the galley proofs for checking.
Enter your text continuously flush left. Do not use any layout functions. Do not split words at the end of a line. **Do not use indentations.** Use a blank line before and a hard return at the end of paragraphs.
Arrange headings flush left. Insert only **one space after punctuation marks.** Insert Greek letters, mathematical symbols etc. with your word-processing program. Use **boldface and italics** as well as sub- and superscript where appropriate.
Tables are part of the text and should be placed at the end of the text file. Separate the individual parts of a table by lines. Use the tabulator instead of spaces. Format operation data on one line in the same column.

Legends to illustrations are part of the text and should also be listed at the end of the text file.

Illustrations

Store illustrations as separate files: do not integrate them in your text. Export **line drawings/vector graphics** as TIF, EPS or WMF format. Use Photoshop for processing and retouching **scanned halftone images.** Save the original scan and the processed version. Export **black and white or color images** as TIF or EPS format in their anticipated size in print.
For detailed information on how to submit illustration data on disk please consult the leaflet or **www.karger.com/disksubmission**

KI 02 DISK

KARGER

Fax +41 61 306 12 34
E-Mail karger@karger.ch
www.karger.com

The Journal Home Page is available at:
www.karger.com/hhe

A stimulating exchange of data and ideas

Journal Publication

Macromolecular Interplay and Traffic on DNA

Molecular Biology, Cell Biology, Genetics

Editors
Paul Bourgine
Alessandra Carbone
Misha Gromov
François Képès

Contents

A detailed understanding of the interplay between regulatory proteins and DNA targets is required to interpret transcriptomic data and to model the dynamics of genetic networks. This special issue deals with two key problems encountered in this respect: One is the control of macromolecular traffic on DNA, the other is the interplay between regulatory proteins acting on the same target gene.

In this issue Dutreix et al. discuss RecA-induced homologous DNA recombination as a paradigm for the search of a site on a long DNA molecule by a smaller molecule. The speed and efficiency of the reaction cannot be explained by brownian motion or molecular tracking, but presumably involves molecular translocations or inchworm local movements. Reinitz et al. present a refined three-layer model of transcriptional regulation. In particular, above direct DNA-binding by factors, the second layer involves adapter molecules binding to DNA-bound factors, while the third layer includes a physical model of how adapters initiate transcription. That second layer is reminiscent of the view presented by Benecke, which he proposes to exploit for genome-wide identification of human DNA targets that are controlled by a combination of certain regulatory factors. The article by Failla et al. introduces two new light microscopy approaches that allow unprecedented spatial resolution. These methods, together with modeling, suggest a relationship between chromosome organization and transcriptional status.

Any life scientist interested in transcriptomics and the dynamics of genetic networks will benefit from the novel views and perspectives presented in this set of papers.

www.complexus.karger.com

KI 03247

Complexus
Editor: Atlan, H. (Paris/Jerusalem)
ISSN 1424–8492
Special issue: Vol. 1, No. 2 (2003)
Included in subscription

Macromolecular Interplay and Traffic on DNA
Editors: Bourgine, P. (Paris); Carbone, A. (Bures); Gromov, M. (Bures); Képès, F. (Evry)
52 p., 19 fig., 5 in color, 2 tab., soft cover, 2003
CHF 29.– / EUR 20.50 / USD 25.25
Prices subject to change
EUR price for Germany, USD price for USA only
ISBN 3–8055–7597–1

KARGER

Order Form

Please send: ____ copy/ies

Postage and handling free with prepayment

Payment:
☐ Check enclosed ☐ Please bill me

Please charge to my credit card
☐ American Express ☐ Diners ☐ Eurocard
☐ MasterCard ☐ Visa

Card No.: ____________________

Exp. date: ____________________

Orders may be placed with any bookshop, subscription agency, directly with the publisher or through a Karger distributor.

Fax: +41 61 306 12 34
S. Karger AG, P.O. Box, CH–4009 Basel (Switzerland)
E-Mail karger@karger.ch, **www.karger.com**

Name/Address: ____________________

Date: ____________________

Signature: ____________________

Human
Heredity

Recent Advances in the Analysis of Genetic Traits

A Celebration of the 30th Anniversary of the Haseman-Elston Method

Editors
Jürg Ott, New York, NY
Derek Gordon, New York, NY

11 figures and 35 tables, 2003

KARGER

Basel · Freiburg · Paris · London · New York ·
Bangalore · Bangkok · Singapore · Tokyo · Sydney

S. Karger
Medical and Scientific Publishers
Basel · Freiburg · Paris · London
New York · Bangalore · Bangkok
Singapore · Tokyo · Sydney

Drug Dosage
The authors and the publisher have exerted every effort to ensure that drug selection and dosage set forth in this text are in accord with current recommendations and practice at the time of publication. However, in view of ongoing research, changes in government regulations, and the constant flow of information relating to drug therapy and drug reactions, the reader is urged to check the package insert for each drug for any change in indications and dosage and for added warnings and precautions. This is particularly important when the recommended agent is a new and/or infrequently employed drug.

P.O. Box, CH–4009 Basel (Switzerland)
Printed in Switzerland on acid-free paper by
Reinhardt Druck, Basel
ISBN 3–8055–7619–6

KARGER

Fax + 41 61 306 12 34
E-Mail karger@karger.ch
www.karger.com

Vol. 55, No. 2–3, 2003

Human Heredity

Contents

KARGER

Fax + 41 61 306 12 34
E-Mail karger@karger.ch
www.karger.com

Access to full text and tables of contents, including tentative ones for forthcoming issues:
www.karger.com/hhe_issues

Human Heredity

Hum Hered 2003;55:78
DOI: 10.1159/000072311

Preface

This special issue of *Human Heredity* is dedicated to the 30th anniversary of one of the seminal papers in genetic epidemiology. In 1972, in their groundbreaking work, Joseph K. Haseman and Robert C. Elston proposed a regression-based parameter-free method for the mapping of quantitative traits in humans (The investigation of linkage between a quantitative trait and a marker locus. Behav Genet 1972;2:3–19). In the ensuing 30 years, their method and its extensions have been widely used for QTL mapping. Recently, with the completion of the genome project and renewed interest in complex human traits, many attempts have been made to improve the power of the Haseman-Elston regression method.

To honor this important anniversary we have invited nine peer-reviewed contributions from respected researchers in the field that in one way or another expand on the framework of the Haseman-Elston method.

More than 10 years ago *Human Heredity* started to celebrate highlights in statistical genetics ('Models and Methods for the Genetic Analysis of Pedigree Data' to celebrate the 20th anniversary of the Elston-Stewart algorithm in 1992, and 'Recent Advances in Genetic Epidemiology', published in honor of Professor Newton Morton's 70th Birthday, published in 2000). With this special issue of *Human Heredity* we continue this tradition.

Jürg Ott, New York, NY

KARGER

Fax +41 61 306 12 34
E-Mail karger@karger.ch
www.karger.com

0001–5652/03/0553–0078$19.50/0

Accessible online at:
www.karger.com/hhe

Original Paper

Human Heredity

Hum Hered 2003;55:79–85
DOI: 10.1159/000072312

Adding Further Power to the Haseman and Elston Method for Detecting Linkage in Larger Sibships: Weighting Sums and Differences

Sanjay Shete[a] Kevin B. Jacobs[b] Robert C. Elston[b]

[a]Department of Epidemiology, University of Texas, M.D. Anderson Cancer Center, Houston, Tex., and
[b]Department of Epidemiology and Biostatistics, Rammelkamp Center for Education and Research, Case Western Reserve University, Cleveland, Ohio, USA

Key Words

Genome scan · Quantitative trait loci · Generalized least squares estimation · Generalized estimating equations · Permutation test

Abstract

Haseman and Elston (H-E) proposed a robust test to detect linkage between a quantitative trait and a genetic marker. In their method the squared sib-pair trait difference is regressed on the estimated proportion of alleles at a locus shared identical by descent by sib pairs. This method has recently been improved by changing the dependent variable from the squared difference to the mean-corrected product of the sib-pair trait values, a significantly positive regression indicating linkage. Because situations arise in which the original test is more powerful, a further improvement of the H-E method occurs when the dependent variable is changed to a weighted average of the squared sib-pair trait difference and the squared sib-pair mean-corrected trait sum. Here we propose an optimal method of performing this weighting for larger sibships, allowing for the correlation between pairs within a sibship. The optimal weights are inversely proportional to the residual variances obtained from the two different regressions based on the squared sib-pair trait differences and the squared sib-pair mean-corrected trait sums, respectively, allowing for correlations among sib pairs. The proposed method is compared with the existing extension of the H-E approach for larger sibships. Control of the type I error probabilities for sibships of any size can be improved by using a generalized estimating equation approach and the robust sandwich estimate of the variance, or a Monte-Carlo permutation test.

Introduction

Many traits of importance to human and medical genetics are quantitative in nature, the phenotype being measured on either a continuous or a discrete scale. Haseman and Elston [1972] (H-E) provided a robust test to detect linkage between a quantitative trait and a genetic marker. H-E regresses the squared sib-pair trait difference on the estimated proportion of marker alleles shared identical by descent (i.b.d.) by the sib pairs. A gain in efficiency can be achieved by performing a generalized least squares (GLS) regression procedure, allowing for the correlations between the pairs of squared sib-pair differences, when we have sibships of size more than two [Single and Finch, 1995].

The squared sib-pair trait difference and the squared sib-pair trait sum are jointly sufficient in the sense that

KARGER

Fax +41 61 306 12 34
E-Mail karger@karger.ch
www.karger.com

0001–5652/03/0553–0079$19.50/0

Accessible online at:
www.karger.com/hhe

Dr. Sanjay Shete
Department of Epidemiology, University of Texas, M.D. Anderson Cancer Center
Houston, TX 77030 (USA)
Tel. +1 713 745 2483, Fax +1 713 792 8261
E-mail sshete@mdanderson.org

with the knowledge of these two quantities the original trait values can be recovered, provided the original trait values are non-negative. H-E ignores the information available in the sib-pair trait sum [Wright, 1997; Gaines and Elston, 1969]. A parametric approach assuming a bivariate normal distribution for the sibs' trait values and then using maximum likelihood has been shown to lead to higher power in certain situations [Fulker and Cherny, 1996; Amos et al., 1997]. Drigalenko [1998] showed that if we use the negative of the squared mean-corrected sib-pair trait sum as a dependent variable in the original H-E method, the resulting regression line is parallel to the original regression line of H-E. Using this observation, Elston et al. [2000] proposed a modification of the original H-E by changing the dependent variable to the mean-corrected product of the sib pairs' trait values. Effectively, in this method equal weight is given to the squared difference and the mean-corrected sum.

This regression approach using both the squared sib-pair trait difference and the mean-corrected sum is often more powerful than using either one of them alone. Situations occur, however, especially when the sibling correlation is large, in which the original H-E method is more powerful than the newly modified method [Palmer et al., 2000]. Xu et al. [2000] reported for independent sib pairs the identical method that we propose here. Forrest [2001] and Visscher and Hopper [2001] have also studied this method for independent sib pairs, mentioning other possible ways of increasing power for this particular situation. For a recent review on this subject see Feingold [2002]. In this communication we develop specifically for larger sibships the proposal by Shete et al. [1999] to use a weighted average of the squared sib-pair trait difference and the squared sib-pair mean-corrected trait sum as the dependent variable in the regression and use generalized least squares (GLS) estimation to obtain a uniformly better procedure for detecting linkage. The weights are inversely proportional to the individual residual variances of each of these regression equations, based on the squared sib-pair trait difference and the squared sib-pair mean-corrected trait sum, respectively. The method proposed by Xu et al. [2002] for larger sibships allows for a possible correlation between the slope estimates obtained from performing separate squared difference and squared sum regressions, but they assume that the sib pairs within a sibship are independent. In particular, they assume that for any two pairs within a sibship the squared differences and squared sums are independent; see equation 4 of Xu et al. [2000]. This assumption is not correct and leads to loss of power. In fact the squared differences (and also the squared sums) between pairs within a sibship are functionally dependent. In our proposed method here we allow for the dependence and perform generalized least squared estimation to obtain an efficient method for detecting linkage. We develop both GLS and general estimating equation (GEE) approaches for arbitrary sibship size, pointing out numerical instabilities that can occur if care is not taken. Based on both theoretical considerations and simulations, the proposed procedures have higher power than competing procedures and we indicate methods of maintaining good control over type I error.

Methods

Generalized Least Squares Regression Procedure Based on the Squared Difference

Let x_{1j} and x_{2j} be the observed trait values for the j-th pair of sibs in a family with s sibs. Let μ be the overall mean. In the original H-E method, the j-th dependent variable is taken to be $(x_{1j} - x_{2j})^2 = [(x_{1j} - \mu) - (x_{2j} - \mu)]^2$. First, for reasons that become clear below, we multiply this dependent variable by $-1/2$ and take the j-th dependent variable based on the difference to be

$$y_j = -\frac{1}{2}[(x_{1j} - \mu) - (x_{2j} - \mu)]^2 = -\frac{1}{2}(x_{1j} - x_{2j})^2. \quad (1)$$

Note that with a sibship of size s, there are $s(s-1)/2$ distinct dependent sib pairs. Despite the obvious non-independence of these pairs, the original H-E retains approximate large sample validity even when all the pairs are assumed to be independent in the linkage analysis [Amos et al. 1989; Collins and Morton, 1995]. It was shown by Blackwelder and Elston [1982] for diallelic markers that there is a slight increase in power when an appropriate correlation structure is used in analyzing sibships of size three. Let s_i, $i = 1,..., k$ be the number of sibs in the i-th family and $n_i = s_i(s_i - 1)/2$ the number of distinct sib pairs. Let

$$n = \sum_{i=1}^{k} n_i$$

and let $\mathbf{y}' = (\mathbf{y}'_1,..., \mathbf{y}'_k)$, where $\mathbf{y}'_i$, the transpose of $\mathbf{y}_i$, is the $1 \times n_i$ vector of dependent variables for the i-th family, $i = 1,..., n$. From Haseman and Elston [1972] we take as our model

$$E(\mathbf{y}) = \mathbf{A}\boldsymbol{\beta}_d, \quad (2)$$

where $\mathbf{A}$ is an $n \times m$ design matrix and $\boldsymbol{\beta}'_d = (\beta_1,..., \beta_m)$ is a vector of parameters. The first column of $\mathbf{A}$ is a column of 1s, corresponding to the intercept β_1, which is of no interest. Following this come one or two columns for each marker locus entered in the above model. Let $\hat{f}_{ij}$ be the conditional probability that the j-th sib pair shares i alleles i.b.d. at a particular marker locus, given all the marker information, and let $\hat{\pi}_j = \hat{f}_{2j} + \hat{f}_{1j}/2$. Then the first of each of these one or two columns is a column whose elements are $\hat{\pi}_j$, and the second (when dominance is believed to be present) is a column whose elements are $\hat{f}_{2j}$. Note that in the original H-E method the regression was performed on $\hat{\pi}_j$ and $\hat{f}_{1j}$. The difference and advantages of regressing on $\hat{\pi}_j$ and $\hat{f}_{2j}$ were discussed by Elston et al. [2000]. Furthermore, because of the coefficient of –1/2 in the dependent variable (1), the elements β_i of $\boldsymbol{\beta}$ are interpretable as locus-specific variance components when $i > 1$ and the markers coincide with trait loci. One can also include in the design matrix columns of products of $\hat{\pi}_j$ and $\hat{f}_{2j}$ to detect the presence of epistasis [Tiwari and Elston, 1997].

The GLS estimator of $\boldsymbol{\beta}_d$ based on model (2) is

$$\hat{\boldsymbol{\beta}}_d = (\mathbf{A}'\mathbf{W}_d^{-1}\mathbf{A})^{-1}\,\mathbf{A}'\mathbf{W}_d^{-1}\mathbf{y} \tag{3}$$

and that of the residual variance is

$$s_d^2 = \frac{\mathbf{y}'\mathbf{W}_d^{-1}[\mathbf{y} - \mathbf{A}\hat{\boldsymbol{\beta}}_d]}{n - m}, \tag{4}$$

where $\mathbf{W}_d$ is the $n \times n$ correlation matrix of $\mathbf{y}$. $\mathbf{W}_d$ is an identity matrix in the case of independent sib pairs, but in general families with multiple sibs a structured correlation matrix should be used [Elston et al., 2000]. For a sibship of size s_i, let $\mathbf{K}_i$ be the correlation matrix for the i-th family. The entries in $\mathbf{K}_i$ are 1, $r_{d,1}$ and $r_{d,0}$, the correlation between elements of $\mathbf{y}$ where the corresponding pairs have both, one, or no sibs in common, respectively. Then $\mathbf{W}_d$ is a block diagonal matrix where the i-th diagonal block is $\mathbf{K}_i$, $i = 1,..., k$. The inverse of $\mathbf{W}_d$, used in the calculation of (3) and (4), can be obtained very rapidly using the results given in the appendix of Elston et al. [2000].

For known values of $r_{d,1}$ and $r_{d,0}$ we can test for the significance of the parameters using usual t test statistics $\hat{\beta}_{d,i}/s_d\sqrt{d_{ii}}$, where $\hat{\beta}_{d,i}$ is the i-th element of $\hat{\boldsymbol{\beta}}_d$, d_{ii} is the i-th diagonal element of $(\mathbf{A}'\mathbf{W}_d^{-1}\mathbf{A})^{-1}$, and s_d is the square root of expression (4). We use a one-sided test because only positive elements of $\hat{\boldsymbol{\beta}}_d$ corresponding to the columns of $\mathbf{A}$ with entries $\hat{\pi}_j$ and $\hat{f}_{2j}$ are considered significant.

Generalized Least Squares Regression Procedure Based on the Squared Mean-Corrected Trait Sum

Analogous to (1), consider a new vector of half the squared sib-pair mean-corrected trait sum

$$z_j = \frac{1}{2}[(x_{1j} - \mu) + (x_{2j} - \mu)]^2. \tag{5}$$

Let $\mathbf{z}' = (\mathbf{z}_1',..., \mathbf{z}_k')$, where $\mathbf{z}_i$ is the dependent variable for the i-th family. Analogous to (2) consider the following model

$$E(\mathbf{z}) = \mathbf{A}\boldsymbol{\beta}_s, \tag{6}$$

where $\mathbf{A}$ is the same design matrix defined earlier and $\boldsymbol{\beta}' = (\beta_1,..., \beta_m)$ is a vector of the same parameters as in model (2) except for the first element β_1, which is of no interest. That is, if we use model (6), then the elements of $\boldsymbol{\beta}$ that correspond to measures of marker allele sharing are the same as in model (2) [Drigalenko, 1998]. In an attempt to combine the information contained in the squared sib-pair trait sums and squared sib-pair trait differences, Elston et al. [2000] considered the dependent variable $(\mathbf{y} + \mathbf{z})/2$, for which the j-th element is the mean-corrected cross-product. That is, they modeled the covariance of the sib pairs directly. This procedure inherently assumes that the information contained in the squared sib-pair trait sum is equal to that in the squared sib-pair trait difference. In other words, modeling the covariance assumes that the residual variances based on regressions of $\mathbf{y}$ and $\mathbf{z}$ are the same. However, these two regressions have, in general, different residual variances.

The generalized least squares estimator of $\boldsymbol{\beta}$ based on model (6) is

$$\hat{\boldsymbol{\beta}}_s = (\mathbf{A}'\mathbf{W}_s^{-1}\mathbf{A})^{-1}\mathbf{A}'\mathbf{W}_s^{-1}\mathbf{z} \tag{7}$$

and that of the residual variance is

$$s_s^2 = \frac{\mathbf{z}'\mathbf{W}_s^{-1}[\mathbf{z} - \mathbf{A}\hat{\boldsymbol{\beta}}_s]}{n - m}, \tag{8}$$

where $\mathbf{W}_s$ is the $n \times n$ correlation matrix of $\mathbf{z}$. Let $\mathbf{L}_i$ be the correlation matrix for the i-th family. The entries in $\mathbf{L}_i$ are 1, $r_{s,1}$ and $r_{s,0}$, the correlation between elements of $\mathbf{z}$ where the corresponding pairs have both, one, or no sibs in common, respectively. Then $\mathbf{W}_s$ is a block diagonal matrix where the i-th diagonal block is $\mathbf{L}_i$, $i = 1,..., k$.

One could perform a usual t test for the significance of a marker measure using the estimates given in (7) and (8), but as before, this is not an efficient procedure. In the next section we describe the method for weighting information contained in the squared sib-pair trait sums and squared sib-pair trait differences.

Generalized Haseman-Elston: Weighting Sums and Differences

Note that the estimated residual variance matrix of $\mathbf{y}$ is $s_s^2\mathbf{W}_d$ and that of $\mathbf{z}$ is $s_s^2\mathbf{W}_s$. With these variances, the new weighted dependent vector we consider is

$$\mathbf{v} = \left(\frac{1}{s_d^2}\mathbf{W}_d^{-1} + \frac{1}{s_s^2}\mathbf{W}_s^{-1}\right)^{-1}\left(\frac{1}{s_d^2}\mathbf{W}_d^{-1}\mathbf{y} + \frac{1}{s_s^2}\mathbf{z}\right). \tag{9}$$

From models (2) and (6) and by the definition of $\mathbf{v}$ in (9), once again we take as our model

$$E(\mathbf{v}) = \mathbf{A}\boldsymbol{\beta}_v, \tag{10}$$

where $\mathbf{A}$ is the design matrix defined earlier and $\boldsymbol{\beta}_v' = (\beta_1,..., \beta_m)$ is a vector of the same parameters as in model (2) or (6) except for the first element β_1, which is of no interest. That is, if we use model (10), then the elements of $\boldsymbol{\beta}$ that correspond to measures of marker allele sharing are the same as in models (2) and (6). The estimated residual variance matrix of $\mathbf{v}$ in (9) can be shown to be

$$\left(\frac{1}{s_d^2}\mathbf{W}_d^{-1} + \frac{1}{s_s^2}\mathbf{W}_s^{-1}\right)^{-1} = \boldsymbol{\Sigma},$$

if $\mathbf{y}$ and $\mathbf{z}$ are uncorrelated. Then the generalized least squares estimator of $\boldsymbol{\beta}$ based on model (10) is

$$\hat{\boldsymbol{\beta}}_v = (\mathbf{A}'\boldsymbol{\Sigma}^{-1}\mathbf{A})^{-1}\mathbf{A}'\boldsymbol{\Sigma}^{-1}\mathbf{v} \tag{11}$$

and that of the residual variance is

$$s_v^2 = \frac{\mathbf{v}'\boldsymbol{\Sigma}^{-1}[\mathbf{v} - \mathbf{A}\hat{\boldsymbol{\beta}}_v]}{n - m}. \tag{12}$$

In general, $\mathbf{y}$ and $\mathbf{z}$ may be correlated. In this case we need to adjust the variance matrix $\boldsymbol{\Sigma}$ given above. The adjusted variance matrix, $\mathbf{D}$ is

$$\boldsymbol{\Sigma} + \frac{1}{s_d s_s}\boldsymbol{\Sigma}(\mathbf{W}_d^{-1}\mathbf{W}_{sd}\mathbf{W}_s^{-1} + \mathbf{W}_s^{-1}\mathbf{W}_{sd}\mathbf{W}_d^{-1})\boldsymbol{\Sigma}, \tag{13}$$

where $\mathbf{W}_{sd}$ is the matrix of correlations between $\mathbf{y}$ and $\mathbf{z}$. Then the GLS estimator of $\boldsymbol{\beta}_v$ based on model (10) is

$$\hat{\boldsymbol{\beta}}_v = (\mathbf{A}'\mathbf{D}^{-1}\mathbf{A})^{-1}\mathbf{A}'\mathbf{D}^{-1}\mathbf{v} \tag{14}$$

and that of the residual variance is

$$s_v^2 = \frac{\mathbf{v}'\mathbf{D}^{-1}[\mathbf{v} - \mathbf{A}\hat{\boldsymbol{\beta}}_v]}{n - m}. \tag{15}$$

To assess significance we again use the t test statistics $\hat{\beta}_{v,i}/s_v\sqrt{m_{ii}}$, where $\hat{\beta}_{v,i}$ is the i-th element of $\hat{\boldsymbol{\beta}}_v$ in (13), m_{ii} is the i-th diagonal element of $(\mathbf{A}'\mathbf{D}^{-1}\mathbf{A})^{-1}$ and s_v is the square root of expression (14). In practice, this model is fitted in two stages. In the first stage, the quantities s_d^2, s_s^2, $\mathbf{W}_d$, $\mathbf{W}_s$ and $\mathbf{W}_{sd}$ are estimated from the available data, and then they are used in the second stage to obtain the estimators of $\hat{\boldsymbol{\beta}}_v$ and s_v^2.

Extension to Generalized Estimating Equations

Further, it is possible to extend this generalized least squares approach to utilize the method of generalized estimating equations (GEE) of Liang and Zeger [1986] by applying (13) iteratively, using $\mathbf{D}$ as the initial working matrix. After each iteration, $\mathbf{W}_d$, $\mathbf{W}_s$, $\mathbf{W}_{sd}$ and $\mathbf{D}$ are updated from the residuals of $\mathbf{y}$ and $\mathbf{z}$ from the previous iteration and new values of s_v^2, and $\hat{\boldsymbol{\beta}}_v$ are computed. In practice this procedure usually converges within 2–4 iterations for most continuous variates and within 3–6 iterations for binary variates coded as (0,1) quantitative variates.

For the above tests to be valid, we have assumed that the correlation matrices $\mathbf{W}_d$, $\mathbf{W}_s$, $\mathbf{W}_{sd}$ are known values. Although these types of tests are fairly robust [Amos et al., 1989; Collins and Morton, 1995], it may be advisable to use the robust sandwich estimator to obtain an estimate of the variance matrix of the parameter estimates that is robust to all forms of misspecification of these correlations [Liang and Zeger, 1986],

$$\mathbf{R}_v = (\mathbf{A}'\mathbf{D}^{-1}\mathbf{A})^{-1}[\mathbf{A}'\mathbf{D}^{-1}(\mathbf{v} - \mathbf{A}\hat{\boldsymbol{\beta}}_v)]'[\mathbf{A}'\mathbf{D}^{-1}(\mathbf{v} - \mathbf{A}\hat{\boldsymbol{\beta}}_v)](\mathbf{A}'\mathbf{D}^{-1}\mathbf{A})^{-1}, \quad (16)$$

and use the same t test statistics $\hat{\beta}_{v,i}/\sqrt{m_{ii}}$ as above, except m_{ii} is now the i-th diagonal element of $\mathbf{R}_v$.

Relationship to the Unified H-E Model

It is useful to notice that the unified H-E model [Xu et al., 2000] is a subset of model (10) obtained by setting $\mathbf{W}_d$, $\mathbf{W}_s$, and $\mathbf{D}$ each to the identity matrix (i.e. ordinary least-squares) and using robust variance estimates (i.e. GEE assuming the identity working matrix). The matrices $\mathbf{W}_d$, $\mathbf{W}_s$, and $\mathbf{D}$ are identity matrices owing to the assumption of an independent working model in the generalized estimating equations. In addition, if one assumes $\mathbf{W}_{sd}$ to be the identity matrix, then for independent sib pairs the unified H-E model is identical to the model of Forrest [2001] and Visscher and Hopper [2001]. Because the method proposed in this paper is for larger sibships, we compare our method to the unified H-E model considered by Xu et al. [2000]. Their estimator is

$$\hat{\boldsymbol{\beta}}_x = (\mathbf{A}'\mathbf{A})^{-1}\mathbf{A}'\mathbf{v} \quad (17)$$

and that of the variance matrix of $\hat{\boldsymbol{\beta}}_x$ is

$$\mathbf{R}_x = (\mathbf{A}'\mathbf{A})^{-1}[\mathbf{A}'(\mathbf{v} - \mathbf{A}\hat{\boldsymbol{\beta}}_x)]'[\mathbf{A}'(\mathbf{v} - \mathbf{A}\hat{\boldsymbol{\beta}}_x)](\mathbf{A}'\mathbf{A})^{-1}. \quad (18)$$

To assess significance in this case we use the t test statistics $\hat{\beta}_{x,i}/\sqrt{m_{ii}}$, where $\hat{\beta}_{x,i}$ is the i-th element of $\hat{\boldsymbol{\beta}}_x$ in (16) and m_{ii} is the i-th diagonal element of $\mathbf{R}_x$.

Estimation of Empirical p Values

For all of the models discussed, we can also estimate an empirical p value of the test statistic using a Monte Carlo permutation procedure with N replicate permutations. For each replicate we use a two-stage permutation procedure: first we permute the allele sharing matrices among sibships of equal size, and second we then permute the allele sharing vectors from the first stage among the pairs within each sibship. The permuting across sibships must be done across sibships of the same size. We permute the allele sharing, rather than the dependent variable, because we want to keep the dependent variable together with all corresponding covariates. After each such permutation we recalculate the test statistic and then determine the proportion of the replicates that are equal to or greater than the statistic calculated from the original observations. We choose N, the number of replicates, such that the estimated empirical p value, $\hat{p}$, is within a proportion w of its true p value, p, with predetermined confidence probability γ. That is, we want the standard error $s_{\hat{p}}$ of $\hat{p}$ to be proportional to $\hat{p}$. This permutation process can be viewed as a set of N independent Bernoulli trials each with success probability p. The sample variance, $s_{\hat{p}}^2$, of $\hat{p}$ is

$$s_{\hat{p}}^2 = \frac{\hat{p}(1-\hat{p})}{N}.$$

So we choose N such that $P(|\hat{p} - p| \leq w\hat{p}) = \gamma$. Using a normal approximation for the distribution of $\hat{p}$, we obtain

$$N = \frac{1-\hat{p}}{w^2\hat{p}}\left[\Phi^{-1}\left(\frac{\gamma+1}{2}\right)\right]^2 \quad (19)$$

where Φ is the standard normal cumulative distribution function. We estimate N by substituting for $\hat{p}$ the p value obtained by assuming the test statistic follows a t distribution, and use this number of replicates to obtain an empirical p value within any prespecified proportion of its true value with a known confidence coefficient. For example, if we wish to estimate a small empirical p-value within 20% of its true value with 95% confidence, then N should be approximately $100/\hat{p}$.

Simulations

We illustrate the performance of the new generalized estimator by performing a limited simulation study to determine the type I error rate and power. Sibship data were simulated for varying sibship sizes under a model in which a codominant (additive) major gene accounts for 20 or 33% of the total trait variability, with 2 alleles having frequencies 20 and 80%. A normally distributed random effect accounts for 30% of the trait variability that is common among siblings (i.e., shared multifactorial components due to common environmental, polygenic inheritance, etc.). The remaining variance is due to an unshared normally distributed random effect for each individual. A marker locus with two equally frequent alleles was simulated to be completely unlinked (in the type I error simulations) or linked (in the power simulations) to the trait locus. Both parental and sibling genotype data were available to estimate the allele sharing i.b.d.

Type I error rates for the generalized H-E model were estimated using the above model with an unlinked marker locus accounting for 20% of the total trait variability. For each configuration, 100,000 simulated replicate data sets were generated, each with 100 nuclear families of fixed

Table 1. Type I error – using the generalized H-E based on the percentage of 100,000 simulated data sets significant at the nominal 5, 1, 0.1, and 0.01% levels

Sibs/ sibship	Sib-ships	Sib pairs	Asymptotic p values				Empirical p values			
			5%	1%	0.1%	0.01%	5%	1%	0.1%	0.01%
2	100	100	5.19	0.98	0.13	0.015	5.00	1.06	0.10	0.015
3	100	300	5.32	1.15	0.11	0.017	5.01	1.05	0.11	0.016
5	100	1,000	5.65	1.68	0.40	0.051	5.03	1.01	0.12	0.021

Table 2. Power – the percentage of simulated data sets significant at the nominal 5, 1, and 0.1% levels

Sibs/ sibship	Sib-ships	Sib pairs	Difference squared H-E			Unified H-E			Generalized H-E		
			5%	1%	0.1%	5%	1%	0.1%	5%	1%	0.1%
2	200	200	12.4	3.3	0.32	18.6	5.4	0.65	18.8	6.1	0.79
3	67	201	17.6	5.3	1.00	16.6	3.8	0.53	19.8	6.4	1.36
4	33	198	22.2	8.4	1.95	15.4	3.8	0.45	23.3	10.9	4.08
5	20	200	22.4	9.9	3.08	14.6	3.2	0.41	26.7	13.3	5.47
6	13	195	22.7	11.9	4.62	12.9	3.5	0.80	25.2	13.5	5.88

sibship size. Significance of each replicate was determined by both asymptotic t test statistics and empirical p values derived from the method described above (with parameters w = 20% and γ = 95%). Table 1 reports the observed type I error rates at the nominal 5, 1, 0.1, and 0.01% levels for sibships of size 2, 3, and 5. For both the asymptotic and empirical tests, in view of the limited number of sibships (100) the observed type I error rates are quite close to their respective nominal significance levels. The error rates are better when estimating empirical p values, though the asymptotic test statistics perform satisfactorily for initial screening and are computationally more practical than using the empirical method to obtain p values. Although there is inflation in the type I error in larger sibships, the additive differences are minuscule. The permutation test we devised here should work in controlling nominal type I error; but if one desires accuracy when the nominal p value is very small, more than 100,000 replicates should be simulated by reducing the w parameter and increasing confidence coefficient parameter γ discussed above. The empirical p values under the null hypothesis were found to be virtually identical to those found for the unified model (17, 18) of Xu et al. [2000] for a variety of simulated genetic models.

Simulations were performed to compare the power of the generalized H-E model (14, 15) to that of the difference squared model (3, 4) and the unified model. The power of the generalized and difference squared models was evaluated using empirically derived cut-points to ensure validity of the results, while the unified model, already utilizing robust variance estimates to ensure validity, used the nominal cut-points. These simulations were performed using the above models with a completely linked marker locus, for a varying number of sibships and sibship sizes. Table 2 reports the observed power for approximately 200 sib pairs at the nominal 5, 1, and 0.1% significance levels for sibships of sizes 2 through 6 when a major additive linked locus accounts for 20% of the total variance. Table 3 reports further results for 300 sib pairs when the major gene accounts for 33% of the trait variability. We note that the generalized H-E method uniformly performs best, though the increase in power over the difference squared for this particular simulation model, while significant, is not dramatic. The gain is small compared with the original H-E because in this simulation model we simulated a residual familial correlation. The lower relative power observed with the unified method is mainly due to the use of robust variance estimates to compensate for the assumption of independence in the working model.

Table 3. Power – the percentage of simulated data sets significant at the nominal 5, 1, and 0.1% levels

Sibs/ sibship	Sib- ships	Sib pairs	Difference squared H-E			Unified H-E			Generalized H-E		
			5%	1%	0.1%	5%	1%	0.1%	5%	1%	0.1%
2	300	300	71.0	45.1	17.3	73.5	51.4	21.3	75.8	50.1	21.3
3	100	300	71.2	46.8	19.8	72.7	41.1	12.2	75.2	52.1	25.0
4	50	300	73.2	49.9	28.7	63.5	29.0	6.5	76.1	57.0	34.7
5	30	300	73.5	54.4	31.4	57.0	22.7	4.3	76.4	59.8	37.8
6	20	300	68.3	52.9	35.5	47.7	18.6	3.6	71.0	57.0	40.2

Discussion

It should be noted that direct computation from the above equations of the GLS and GEE estimators, and of the robust variance estimator, is ill-advised because of potentially extreme numerical instability when using conventional floating-point arithmetic [Golub and Van Loan 1996; Björck 1996]. A numerically stable algorithm using orthogonal transformations was suggested by Paige [1979], but is computationally impractical for this application because it destroys the block diagonal structure of **A** and **D**. However, in practice it is usually sufficient to use the methods described in Golub [1965] and Björck [1996] that are numerically stable for standard univariate linear least squares. The results will then be numerically stable so long as the Cholesky triangle of **D** is not too ill-conditioned and care is taken to avoid ill-conditioned intermediate steps in the computations.

In practice, we also recommend restricting the estimates $r_{d,1}$, $r_{d,0}$, $r_{s,1}$, and $r_{s,0}$ to be non-negative, and for this reason the sandwich estimator (16) of the variance matrix or estimation of empirical p values may be necessary to control type I error. The three correlations that occur in $\mathbf{W}_{sd}$ need not be restricted unless the numerical methods used require $\mathbf{W}_{sd}$ to be positive semi-definite.

In conclusion, we have indicated how weighting the squared sib-pair trait difference and the squared sib-pair mean-corrected trait sum, and modeling the sibling correlations using GLS and GEE, can be used to increase the power of the H-E regression procedure for sibships of arbitrary size, and yet maintain control of the probability of Type I error. All of the various models discussed are implemented in the latest version of the computer program SIBPAL [S.A.G.E., 2003] and we are currently investigating the circumstances in which each is the most powerful. The results reported here for the generalized H-E used option W4 in SIBPAL, and those for the unified H-E used option W2, the robust variance estimator, and the independence working matrix options.

Acknowledgements

This work was supported in part by US Public Health Service research grant GM 28365 from the National Institute of General Medical Services, and by resource grant RR 03655 from the National Center for Research Resources.

References

Amos CI, Elston RC, Wilson AF, Bailey-Wilson JE: A more powerful robust sib-pair test of linkage for quantitative traits. Genet Epidemiol 1989; 6:435–450.

Amos CI, Krushkal J, Thiel TJ, Young A, Zhu DK, Boerwinkle E, Andrade M: Comparison of model-free linkage mapping strategies for the study of complex trait. Genet Epidemiol 1997; 14:743–748.

Björck A: Numerical methods for least squares problems. Society for Industrial and Applied Mathematics, Philadelphia, 1996; section 2.8.3.

Blackwelder WC, Elston RC: Power and robustness of sib-pair linkage tests and extension to larger sibships. Commun Stat Theor Meth 1982;11: 449–484.

Collins A, Morton NE: Nonparametric tests for linkage with dependent sib pairs. Hum Hered 1995;47:333–353.

Drigalenko E: How sib-pairs reveal linkage. Am J Hum Genet 1998;63:1243–1245

Elston RC, Buxbaum S, Jacobs KB, Olson JM: Haseman and Elston revisited. Genet Epidemiol 2000;19:1–17.

Feingold E: Regression-based quantitative-trait-locus mapping in the 21st century. Am J Hum Genet 2002;71:217–222.

Forrest WF: Weighting improves the 'New Haseman-Elston' method. Hum Hered 2001;52:47–54.

Fulker DW, Cherny SS: An improved multipoint sib-pair analysis of quantitative traits. Behav Genet 1996;26:527–532.

Gaines RE, Elston RC: On the probability that a twin pair is monozygotic. Am J Hum Genet 1969;21:457–465.

Golub GH: Numerical methods for solving least squares problems. Numer Math 1965;7:206–216.

Golub GH, Van Loan CF: Matrix computations. Baltimore, Johns Hopkins University Press, 1996.

Haseman JK, Elston RC: The investigation of linkage between a quantitative trait and a marker locus. Behav Genet 1972;2:3–19.

Liang KY, Zeger SL: Longitudinal data analysis using generalized linear models. Biometrika 1986;73:13–22.

Paige CC: Numerically stable computations for general univariate linear models. Comm Statist 1979;7:437–453.

Palmer LJ, Jacobs KB, Elston RC: Haseman and Elston revisited: The effects of ascertainment and residual familial correlations on power to detect linkage. Genet Epidemiol 2000;19:456–460.

S.A.G.E.: Statistical Analysis for Genetic Epidemiology, Release 4.4. Computer program package available from Statistical Solutions Ltd, Cork, Ireland, 2002.

Shete S, Jacobs KJ, Elston RC: Adding further power to the Haseman and Elston method for detecting linkage. Genet Epidemiol 1999;17: 194.

Single RM, Finch SJ: Gain in efficiency from using generalized least squares in the Haseman-Elston test. Genet Epidemiol 1995;12:889–894.

Tiwari HK, Elston RC: Linkage of multilocus components of variance to polymorphic markers. Ann Hum Genet 1997;61:253–261.

Visscher PM, Hopper JL: Power of regression and maximum likelihood methods to map QTL from sib-pair and DZ twin data. Ann Hum Genet 2001;65:583–601.

Wright FA: The phenotypic difference discards sib-pair QTL linkage information. Am J Hum Genet 1997;60:740–742.

Xu X, Weiss S, Wei LJ:A unified Haseman-Elston method for testing linkage with quantitative traits. Am J Hum Genet 2000;67:1025–1028.

Original Paper

Human Heredity

Hum Hered 2003;55:86–96
DOI: 10.1159/000072313

Regression Models for Linkage: Issues of Traits, Covariates, Heterogeneity, and Interaction

Daniel J. Schaid[a] Jane M. Olson[b] W. James Gauderman[c]
Robert C. Elston[b]

[a]Department of Health Sciences Research, Division of Biostatistics, Mayo Clinic/Foundation, Rochester, Minn.,
[b]Department of Epidemiology and Biostatistics, Case Western Reserve University, Cleveland, Ohio, and
[c]Department of Preventive Medicine, Division of Biostatistics, University of Southern California,
Los Angeles, Calif., USA

Key Words

Haseman and Elston · Linear regression · Logistic regression · Quantitative trait loci · Sib-pairs

Abstract

Regression methods offer a common framework to analyze linkage for quantitative trait loci as well as linkage for affection status using affected sib-pairs. Although numerous papers on regression methods for linkage have been published, some common themes and important caveats tend to be scattered across the literature. For example, the typical approach is to regress a function of traits on identical-by-descent (IBD) information, but the reversal (regression of IBD on a function of traits) offers important insights. A second example is the use of regression equations to assess linkage heterogeneity or gene-environment interaction, and why these two different etiologies are difficult to distinguish with affected sib-pair data. A third example has to do with the differences, and similarities, between linear regression and nonlinear regression methods for affected sib-pair data. The purposes of this paper are to review some recent developments in the linkage regression framework, to emphasize strengths and weaknesses of various proposed methods, and to highlight some important assumptions and caveats.

Introduction

Regression methods offer a common framework to analyze linkage of quantitative trait loci and linkage of affection status based on affected sib-pair data. The purposes of this paper are to review some recent developments in this area, in order to provide a common framework for regression-based linkage analyses, to emphasize strengths and weaknesses of various proposed methods, and to highlight some important caveats.

There has been a flurry of papers published on regression methods for linkage analysis of quantitative traits, many of which extend and enhance the method proposed by Haseman and Elston [1]. Some of these newly proposed regression methods are close competitors to variance component linkage analysis [2], and even asymptotically equivalent to variance component analysis [3, 4]. Although most of the regression-based methods regress a

KARGER

Fax +41 61 306 12 34
E-Mail karger@karger.ch
www.karger.com

0001–5652/03/0553–0086$19.50/0

Accessible online at:
www.karger.com/hhe

Daniel J. Schaid, Ph.D.
Harwick 775, Division of Biostatistics, Mayo Clinic/Foundation
200 First ST SW
Rochester, MN 55905 (USA)
Tel. +1 507 284 0639, Fax +1 507 284 9542, E-Mail schaid@mayo.edu

function of the traits of relative pairs on the proportion of alleles shared identical-by-descent (IBD) at marker loci, new developments by Sham et al. do just the opposite: regress the proportion of alleles shared IBD on functions of the traits of relative pairs [5]. We review the advantages and disadvantages of this approach, but see also Feingold [6] for an informative review and commentary on these developments.

For linkage analysis of affection status with affected sib-pairs (ASPs), a number of regression-based methods have been developed in order to account for covariates that influence either linkage heterogeneity or gene-environment interaction. Some of the proposed methods are based on linear regression, and some on non-linear regression via logistic models. Although the methods appear disparate, the regression methods offer a common framework to consider the similarities, strengths, and weaknesses of the various techniques.

Table 1. Definitions of dependent Y for regression on $\hat{\pi}$*

Y	β
$D^2 = (y_1 - y_2)^2$	$-2\sigma_g^2$
$S^2 = (y_1 + y_2)^2$	$+2\sigma_g^2$
$CP = y_1 y_2$	σ_g^2
$W = S^2/(1+\rho)^2 - D^2/(1-\rho)^2$	σ_g^2

* Assume trait y is mean-corrected, by subtracting the mean from the trait.
ρ = Residual sib correlation, after acounting for genetic component.
σ_g^2 = Genetic variance component.

Regression Models for Quantitative Traits

To simplify the presentation of regression methods for quantitative traits, we initially focus on sib-pairs, and later briefly outline methods for other types of relative pairs. Let Y be a function of the traits y_1 and y_2 of the members of a sib-pair. This function could be the squared difference, the cross-product (after subtracting the mean from each member's trait), or other functions; see examples in table 1. Discussion on choice of the Y function has been a focus of several papers [3, 4, 7–11], and succinctly summarized by Feingold [6, 12].

The usual approach to test for linkage between a quantitative trait and a genetic marker is to regress Y on $\hat{\pi}$ according to the regression equation $E[Y|\hat{\pi}] = \alpha + \beta\hat{\pi}$, where $E[Y|\hat{\pi}]$ denotes the conditional expectation of Y, given $\hat{\pi}$, and $\hat{\pi}$ is the estimated proportion of alleles shared IBD at the genetic marker locus. Because the true proportion of alleles shared IBD, denoted π, is often not directly observed, it must be estimated by use of the marker data. This proportion can be estimated as $\hat{\pi} = \hat{z}_2 + 0.5\hat{z}_1$, where $\hat{z}_i$ is the estimated probability that a sib pair shares i alleles IBD at the marker locus (note that the 'hat' is used to denote parameter estimates). Recent work has shown that when Y is a weighted average of the squared difference and squared sum (see W in table 1) [4, 5, 11], this linear regression is asymptotically equivalent to variance component methods for quantitative trait linkage analysis based on the assumption of multivariate normality.

A few key points regarding the original HE method (with Y defined as the squared-trait-difference [1]) vs. the revised HE method (with Y defined as the mean-corrected cross-product [8]) are as follows: (1) the original HE can be more powerful than the revised HE when there is strong ascertainment on the trait values (for instance, when the trait values of the sibs are above the 70th percentile of the population distribution of the trait [13]), or when the residual sib correlation – after accounting for genetic components of variance – is high (for instance, when residual sib correlation >0.27 [3, 13]); (2) the most efficient way to combine the original and revised HE regressions is by a weighted sum of the squared difference and the squared mean-corrected sum [4]; (3) for a random sample, correction of the cross-product by the population mean can be more powerful than correction by the sample mean, but mis-specification of the population mean can lead to a large loss in power [2]. In the case of a selected sample, the 'best' value to use for the mean-correction, with 'best' meaning little loss in power, lies between the population mean and the sample mean [13].

For our presentation, it is important to recognize that the t statistic for testing linkage (H_0: $\beta = 0$) does not change if we interchange Y and $\hat{\pi}$ in the regression (see Appendix). Furthermore, the lod score is also invariant, because lod score = $t^2/[2\,ln(10)]$. Hence, for calculation of lod scores, it doesn't matter which is the dependent vs. the independent regression term. Another way to think about the equivalency between regressing Y on $\hat{\pi}$ or $\hat{\pi}$ on Y (using linear regression) is that the test of H_0: $\beta = 0$ from linear regression is equivalent to testing no correlation of Y with $\hat{\pi}$ using the t test (H_0: $\rho = 0$, tested by $\hat{\rho}$, Pearson's correlation coefficient). The calculation and interpretation of $\hat{\rho}$ does not depend on whether Y or $\hat{\pi}$ is declared as the 'dependent' variable. The interesting thing about this is that it shows that bivariate normality is not necessary for the t distribution to be valid when testing a correlation – it is sufficient to have Y normally distributed given $\hat{\pi}$, or $\hat{\pi}$ normally distributed given Y. But this matters little in large samples, in which the regression coefficient, being a weighted average, becomes normally distributed by the central limit theorem.

The interchange of Y and $\hat{\pi}$ extends to regression models that include adjustment for pair-specific covariates. For example, if X is a pair-specific covariate, such as the difference in covariate values between members of a pair, then equivalent tests of significance for β will be obtained from either of the following models:

$$E[Y|\hat{\pi}] = \alpha + \beta\hat{\pi} + \gamma X \quad \text{or} \quad E[\hat{\pi}|Y] = \alpha + \beta Y + \gamma X.$$

The test of whether $\beta = 0$ is based on the partial correlation between Y and $\hat{\pi}$, given X, which does not depend on which is declared the dependent variable. Of course, the estimates and interpretations of the regression coefficients depend on whether Y or $\hat{\pi}$ is the dependent variable. In fact, Sham et al. [5] note that the regression coefficients will be biased when sampling is through the dependent variable. It is common to sample subjects according to their trait values, but unusual to sample subjects according to their marker genotypes. Hence, to avoid biased estimators of regression coefficients, in terms of how the slope parameter captures variation attributed to genetic effects, Sham et al. regress $\hat{\pi}$ on both $D^2 = (y_1 - y_2)^2$ and $S^2 = (y_1 + y_2)^2$. From these points, it should be clear that for the calculation of lod scores, how the Y value is computed from the traits is critical, but whether Y is on the left-hand side or right-hand side of the regression equation is not important.

Person-Specific vs. Pair-Specific Covariates

Person-specific covariates are typically included in the analysis in order to reduce the residual error of the model, in order to improve the linkage signal. In contrast, pair-specific covariates allow evaluation of differences between pairs, rather than the influence of individuals, which is critical to evaluate whether the linkage signal varies with the pair-specific covariate values, and hence allows evaluation of linkage heterogeneity or gene-environment interaction. A subtle distinction between variance component linkage analysis vs. the HE regression-based methods is how covariates are handled.

Variance component analysis directly models person-specific covariates, and the parameters for covariates are estimated simultaneously with the parameters that depend on linkage. The variance component model can also incorporate pair-specific covariates, to evaluate gene-covariate interaction, by including the interaction of IBD and covariate in the covariance matrix [14].

In contrast, modeling of person-specific covariates using HE regression methods would normally require a two-step regression approach. First, the trait can be regressed on covariates (ignoring the linkage information), and then the residuals from this first regression can be treated as new 'traits' in a second-step HE linkage regression. Limitations of this two-step procedure are that the parameters for covariates and those for linkage are not estimated simultaneously, and gene-environment interactions cannot be modeled. However, interactions are easily modeled by pair-specific covariates in the HE regression, and because this method directly models pair-specific covariates, they can be flexibly modeled, such as by polynomial regression. Note that pair-specific covariates either (1) can be formed from person-specific covariates as described above, or (2) may represent variables that are truly pair-specific, such as maternal genotype or family history of disease. The efficiency of variance component analysis relative to HE regression for estimating covariate effects, testing linkage after adjusting for covariates, and assessing linkage heterogeneity or gene-environment interaction is an area requiring further study.

Linkage Heterogeneity and Gene-Environment Interaction

It is important to consider how to evaluate and interpret linkage heterogeneity and gene-environmental interaction in the regression framework. For example, Gauderman et al. [15] proposed a modification of the HE regression model for quantitative trait analysis to allow for gene-environment interaction. For an assumed definition of Y and appropriate X that quantifies some function of the sib-pair covariates, they proposed the model

$$E[Y|X,\hat{\pi}] = \alpha + \beta_e X + \beta_g\hat{\pi} + \beta_{ge}\hat{\pi}X. \tag{1}$$

If one were interested in testing no gene-environment interaction (i.e., H_0: $\beta_{ge} = 0$), adjusting for the presence of linkage (i.e., allowing $\beta_g \neq 0$), then one could switch the roles of Y and $\hat{\pi}$ in expression (1) without affecting the resulting statistical test. In contrast, if the focus is a test of the joint null hypothesis of no linkage and no gene-environment interaction (H_0: $\beta_g = \beta_{ge} = 0$), in an effort to use interaction information to increase power to detect linkage, then it is not valid to simply interchange Y with $\hat{\pi}$, because $\hat{\pi}$ occurs twice on the right hand side, and the hypothesis test involves both β_g and β_{ge}. We cannot interchange Y and $\hat{\pi}$ when hypothesis tests focus on more than one regression coefficient.

Now suppose that there is no interaction, so that $\beta_{ge} = 0$ in expression (1). Should the X covariate still be included? The answer is yes, if $\hat{\beta}_e$ differs from zero, because in this case including the covariate X reduces the residual sum of squares, and thus, unless the number of degrees of freedom is small, increases power; the linkage, in this case, is not 'dependent on' the covariate, and valid inferences would be the same if Y and $\hat{\pi}$ are interchanged. Inclusion of X as a covariate may also help induce normality of the residuals.

The regression equation (1) is general, applying also to binary affection status, which can be represented as a quantitative trait. Without loss of generality, we assume that y has values of 1 for affected and 0 for unaffected persons, but any two different values could be used to code affection status. First consider the special case when there is no X covariate effect, $\beta_e = \beta_{ge} = 0$, and define $Y = (y_1 - y_2)^2$. When sib-pairs are concordant with respect to their affection status, $Y = 0$, and when discordant, $Y = 1$. Then, testing H_0: $\beta_g = 0$ is the same as testing the equality of $\hat{\pi}$ between concordant and discordant pairs. Now consider the full regression equation (1) that allows for gene-environment interaction, and define Y the same way, but switch Y and $\hat{\pi}$:

$$E[\hat{\pi}|X,Y] = \alpha + \beta_e X + \beta_g Y + \beta_{ge} YX. \qquad (2)$$

Substituting 0 or 1 for Y into expression (2) results in

$$E[\hat{\pi}|X,Y] = \begin{cases} \alpha + \beta_e X & \text{if concordant} \\ (\alpha + \beta_g) + (\beta_e + \beta_{ge})X & \text{if discordant} \end{cases} \qquad (3)$$

This illustrates that β_g measures whether the allele sharing differs between concordant vs. discordant sib pairs (as it should if there were linkage), and that β_{ge} measures whether the dependence of allele sharing on covariate X differs between concordant vs. discordant sib pairs. Hence, equal slopes between concordant and discordant sib pairs implies no gene-environment interaction ($\beta_{ge} = 0$), and equal intercepts implies no linkage ($\beta_g = 0$); these hypotheses can be easily tested in the regression framework.

Regression Models for ASP Identity-by-Descent Probabilities and Covariates

If the IBD sharing probabilities (π) for ASPs differ in a systematic way according to covariates measured on the ASPs, then the linkage signal can be increased by using this information. A number of regression-based methods for model-free analysis of ASPs have been proposed, and we review them as two separate categories: linear vs. nonlinear regression.

Linear Regression Models

Linear regression for ASPs is an extension of the mean sharing statistic [16]; the mean sharing statistic has the form $Z = (\bar{\pi} - 0.5)/SE(\bar{\pi})$, where $\bar{\pi}$ and $SE(\bar{\pi})$ are the sample mean and its standard error, respectively, for the $\hat{\pi}_i$ values in a sample of affected sib pairs. This model can also be written as

$$E[\hat{\pi}] = \alpha,$$

with the null hypothesis of no linkage, H_0: $\alpha = 0.5$, tested by a t statistic. For the regression extension, the dependent variable, $\hat{\pi}$, is regressed on covariate X, in the usual linear regression model,

$$E[\hat{\pi}|X] = \alpha + \beta_e X. \qquad (4)$$

This expression (4) can be derived from the more general expression (3) by Y = 0 setting for all ASPs (e.g., all pairs are concordant affected). A test of linkage can be obtained either by testing H_0: $\alpha = 0.5$ (provided X has been centered on its sample mean), or by testing the joint null hypothesis H_0: $\alpha = 0.5$, $\beta_e = 0$. A test of heterogeneity alone is equivalent to testing whether $\beta_e \neq 0$. Note that heterogeneity in this context is not restricted to 'linkage heterogeneity'; rather, heterogeneity in this context means that the IBD sharing varies according to the value of X.

The reasons why IBD sharing could depend on covariate X are varied, and subject to the specific details of any given study. To emphasize this point, consider linkage heterogeneity vs. gene-environment interaction. Linkage heterogeneity occurs when some ASPs are linked and others are not linked. Gene-environment interaction implies that the penetrance of the gene varies with an environmental risk factor such that the magnitude of π varies over the level of sib-similarity of environmental risk factor measured by the X value. It might be possible to infer linkage heterogeneity if one could stratify on the levels of the covariate, and show that within some strata there is no evidence of linkage (i.e., $\bar{\pi} = 0.5$), yet other strata show evidence favoring linkage ($\bar{\pi} > 0.5$). However, whether this represents linkage heterogeneity or gene-environment interaction depends on the context of the study. For example, in a study of ASPs with lung cancer and tobacco smoking behavior as a covariate, if $X = 0, 1, 2$ for ASPs that are concordant for non-smoking, discordant for smoking, or concordant for smoking, then finding $\bar{\pi} = 0.5$ among the concordant non-smokers ($X = 0$) and $\bar{\pi} > 0.5$ within the other strata ($X > 0$) might imply that the gene is expressed only in the presence of tobacco smoke, an example of gene-environment interaction. Statistically, it can be difficult to distinguish whether some subsets are truly not linked, vs. whether all subsets are linked but with $\bar{\pi}$ varying over the subsets due to gene-environment interaction. None the less, as shown by Gauderman and Siegmund [17], use of regression model (4) can increase the power to detect linkage when gene-environment interaction is at least moderate. For further discussion of the advantages of linear regression in the presence of gene-environment interaction, see Gauderman and Siegmund [17].

Although we may not be able to distinguish between linkage heterogeneity vs. gene-environment interaction for ASP designs, the magnitude of variation in $\hat{\pi}$ caused by linkage heterogeneity is likely to be greater than variation in $\hat{\pi}$ caused by gene-environment interaction. A covariate such as a measure of clinical disease could in principle fully separate linked from unlinked subgroups. If this is the case, then the difference in the proportion of alleles shared IBD among subgroups should be greater when some subgroups are not linked and some are linked (i.e., linkage heterogeneity) than when all subgroups are linked yet the susceptibility gene interacts with the covariate to cause the amount of allele sharing to differ over the subgroups (i.e., gene-environment interaction). In any event, quantifying the expected magnitude of variation among covariate-defined subgroups is critical when designing studies to evaluate either linkage heterogeneity or gene-environment interaction.

For ASP studies, affection status is often considered the primary variable to evaluate for linkage, and other clinical traits (either discrete or continuous) are treated as covariates, in order to evaluate whether linkage of affection status varies over the other clinical traits. But, what if affection status is not linked, yet one of the other clinical traits is linked? How can we define covariates to distinguish whether a clinical trait explains heterogeneity in linkage for affection status, vs. linkage of the clinical trait? A good example that illustrates this distinction is provided by prostate cancer affected sib-pair studies that measure Gleason grade. Gleason grade is based on the architectural pattern of biopsy or prostatectomy specimens, and is measured by adding two histological pattern grades, each of which is individually scored from 1 to 5 (total score range is 2 to 10). A high score is generally associated with poor prognosis. Goddard et al. [18] used the mean grade for a sib-pair, $(y_1 + y_2)/2$, as a covariate to explain heterogeneity of IBD sharing. In contrast, Witte et al. [19] treated Gleason score as a quantitative trait and used the original HE, $(y_1 - y_2)^2$, and the revised HE methods to evaluate linkage. As shown by Haseman and Elston [1], when the quantitative trait y is linked, $E[(y_1 - y_2)^2 | \hat{\pi}] = \alpha + \beta\hat{\pi}$, where β depends on the genetic variance. Following the same procedures as Haseman and Elston (see table 1 in [1]), it can be shown that $E[(y_1 + y_2)/2 | \hat{\pi}]$ does not depend on $\hat{\pi}$, nor does $E[(y_1 - y_2) | \hat{\pi}]$, even if the trait y is linked. By interchanging Y and $\hat{\pi}$ in the regression equation (so that the Y is on the right-hand-side), it follows that using either the sum $(y_1 + y_2)$ or the difference $(y_1 - y_2)$ as a covariate is a way to measure 'pure heterogeneity' in IBD sharing among ASPs, because it is not confounded by the possible linkage of the clinical trait. In contrast, use of $(y_1 - y_2)^2$ as a covariate in a heterogeneity analysis can lead to confounding, because we cannot be sure whether y explains heterogeneity in linkage of affection status, or whether y itself is linked.

Many studies sample ASPs without unaffected subjects, which make some implicit assumptions that are important to consider. A basic difference between the ASP method vs. sampling that includes unaffected subjects is that the latter is in a sense 'controlled'. Consider a sample of ASPs and discordant sib pairs (DSPs). In the regression method the two kinds of pairs are contrasted with each other, as emphasized in the discussion just prior to expression (2), as well as by expression (3). In contrast, when only ASPs are sampled, they are each contrasted with null Mendelian expectation, because the regression equation (without the X covariate) reduces to a model with only the intercept, $E[\hat{\pi}] = \alpha$. The t statistic to test for linkage ($\alpha > 0.5$) is equivalent to the mean sharing statistic. Hence, when sampling only ASPs, it must be assumed that without linkage $\alpha = 0.5$. When the penetrance of the susceptibility gene is low, so that allele sharing among unaffected subjects is not far from the null expectation, sampling only ASPs is expected to give more power than sampling both ASPs and DSPs, but at the expense of assuming Mendelian segregation and/or (if required) correct marker allele frequencies. An ASP analysis could be fooled by selection (and there is a lot of selection between conception and birth) which could cause all pairs (ASPs

Table 2. Logistic regression models for IBD sharing probabilities and covariates

Model	$z_0(X;\beta)$	$z_1(X;\beta)$	$z_2(X;\beta)$	References
Trinomial	$\frac{1}{1+2e^{\beta_{1.0}+\beta_{1.1}X}+e^{\beta_{2.0}+\beta_{2.1}X}}$	$\frac{2e^{\beta_{1.0}+\beta_{1.1}X}}{1+2e^{\beta_{1.0}+\beta_{1.1}X}+e^{\beta_{2.0}+\beta_{2.1}X}}$	$\frac{e^{\beta_{2.0}+\beta_{2.1}X}}{1+2e^{\beta_{1.0}+\beta_{1.1}X}+e^{\beta_{2.0}+\beta_{2.1}X}}$	22, 24
Logistic – Multiplicative	$\frac{1}{(1+e^{\beta_0+\beta_1X})^2}$	$\frac{2e^{\beta_0+\beta_1X}}{(1+e^{\beta_0+\beta_1X})^2}$	$\frac{(e^{\beta_0+\beta_1X})^2}{(1+e^{\beta_0+\beta_1X})^2}$	2, 26–28
Logistic – Minimax	$\frac{1}{-1.634+5.634e^{\beta_0+\beta_1X}}$	$\frac{2e^{\beta_0+\beta_1X}}{-1.634+5.634e^{\beta_0+\beta_1X}}$	$\frac{3.634e^{\beta_0+\beta_1X}-2.634}{-1.634+5.634e^{\beta_0+\beta_1X}}$	18
	$\frac{e^{\beta_0+\beta_1X}}{1.5504+2.45e^{\beta_0+\beta_1X}}$	$\frac{0.5504+1.45e^{\beta_0+\beta_1X}}{1.5504+2.45e^{\beta_0+\beta_1X}}$	$\frac{1}{1.5504+2.45e^{\beta_0+\beta_1X}}$	24

and DSPs) to share more marker alleles IBD than expected. How often this occurs when using microsatellite markers is not clear, because it is unusual for investigators to sample unaffected subjects, and unusual to test Mendelian transmission. Selection will depend on whether a microsatellite marker is in linkage disequilibrium with a gene that is highly selected. As studies begin to incorporate more densely spaced markers, such as single nucleotide polymorphisms (SNPs), the possibility for selection of a marker seems more likely.

Now, what exactly does the slope β_e in expression (4) measure? This regression slope is influenced by the correlation between $\hat{\pi}$ and X, because $\beta_e = Corr(\hat{\pi}, X)\sqrt{Var(\hat{\pi})/Var(X)}$. If we assume that in the general population there is no correlation of the susceptibility gene with environmental risk factor, so that at the population level $Corr(\hat{\pi}, X) = 0$, then the estimate $\hat{\beta}_e$ can be used to construct a valid t statistic to test the association between $\hat{\pi}$ and X in a sample of ASPs. However, if there is a population correlation of the susceptibility gene with the environmental risk factor, and only ASPs are sampled, then the slope estimate $\hat{\beta}_e$ can be biased, causing false inference of either linkage heterogeneity or gene-environment interaction. This situation is analogous to evaluating a gene-environment interaction with a series of cases (without controls), and measuring interaction by the correlation of gene frequency with environmental exposure. This 'case-only' design requires the assumption that the gene and environmental risk factor are independent in the general population [20].

In contrast, when DSPs are also sampled, expression (3) illustrates that the DSPs serve as a 'control', to measure the difference in slopes between DSPs and ASPs, so that the test for gene-environment interaction in this case (i.e., based on the estimate of β_{ge} of expression 3) is not biased by population association of a gene with the environmental risk factor. A situation in which a gene could be correlated with an environmental risk factor occurs when both of these factors are correlated with a third factor, such as ethnic background (e.g., population stratification).

Nonlinear Regression Models

Several authors have extended the popular model-free maximum lod score (MLS) of Risch [21] by allowing the IBD sharing probabilities for an ASP to depend on covariate information [22–28]; for a comparison of several methods, see Zhang et al. [2]. The basis for these models is that the three IBD sharing probabilities, z_i (i = 0, 1 2), represent a trinomial distribution, and the trinomial probabilities can be modeled by a form of the logit transform of the z_i values, much like logistic regression. A summary of proposed models is given in table 2. The most general model is the trinomial for the three IBD categories. When IBD = 0 is chosen as the reference category, the trinomial logit model can be expressed as two logit models, with coefficients for the logit of IBD = 1 vs. IBD = 0 ($\beta_{1,0}$, $\beta_{1,1}$), and coefficients for IBD = 2 vs. IBD = 0 ($\beta_{2,0}$, $\beta_{2,1}$). However, for p covariates (including the intercept), there are $2p$ coefficients to estimate, which can reduce power to detect linkage heterogeneity.

To reduce the number of regression parameters, a number of alternative models have been proposed (see [24] for more thorough discussion). One approach, denoted 'Logistic – Multiplicative' in table 2, uses the usual binary logistic regression to model the event $Y = 1$ if an allele is transmitted IBD from a parent to both members of an ASP (and $Y = 0$ otherwise). This model then multiplies the logistic probabilities for the two parents of an ASP, implicitly assuming that the IBD transmissions are independent between the two parents. Another approach, denoted 'Logistic – Minimax' imposes the minimax constraint proposed by Whittemore and Tu [29], $z_1 = 0.335 + 0.58z_0$, which minimizes the maximum percentage increase in sample size required when using a sub-optimal test among those tests that constrain the sharing probabilities to the genetically-possible triangle [30]. Although the forms of the two minimax regression models in table 2 are different, it is possible to map from one model to the other (i.e., one model is a reparameterization of the other). For example, the following steps can be used to convert from the model proposed by Greenwood and Bull [24] to that proposed by Goddard et al. [18]: multiply the z_i values for the model of Greenwood and Bull by $e^{-(\beta_0+\beta_1 X)}$; replace $-\beta_0$ with α_0 and $-\beta_1$ with α_1; set the z_0 values equal to each other for the two models to find that $e^{\alpha_0+\alpha_1 X}$ for the Greenwood and Bull model is equal to $3.635e^{\beta_0+\beta_1 X} - 2.635$) for the model by Goddard et al. Hence, although the estimated regression coefficients may differ between the two models, the likelihood ratio statistics and lod scores for testing the influence of covariates will be the same.

Although the logistic regression models have focused on ASPs, it is possible to include both ASPs and DSPs (as well as concordant unaffected pairs) by using a covariate that indicates concordance status. This is equivalent to analyzing ASPs and DSPs separately (contrasting each with Mendelian segregation of markers), and summing the lod scores. But, because the contrast between ASPs and DSPs is measured by a covariate parameter, it may be more robust to departure from Mendelian segregation of marker alleles to simply test for linkage by evaluating whether this covariate parameter departs from zero. More work along these lines is needed to evaluate how best to model mixtures of different types of pairs.

Linear vs. Nonlinear Regression

When the regression effect is not large, the nonlinear regression models for the expected proportion of alleles shared IBD are not that different from the linear regression model. To see this, consider the multiplicative logistic model. After substituting z_2 and z_1 from table 2 into $\pi = z_2 + 0.5z_1$, and taking a first-order Taylor series approximation around $\beta_1 = 0$, it can be shown that $\pi(X;\beta_0, \beta_1) \approx \pi(\beta_0) + \gamma X$, where $\pi(\beta_0) = e^{\beta_0}/[1 + e^{\beta_0}]$, the proportion of alleles shared IBD without a covariate effect, and $\gamma = \beta_1\pi(\beta_0)[1 - \pi(\beta_0)]$. That is, the regression slope for linear regression, γ, depends on the corresponding parameter for the logistic model (β_1) as well as the intercept $\pi(\beta_0)$. This really doesn't matter, as there are still only two parameters for each type of model. In contrast, if the effect of X is not well approximated by this first-order approximation, and a second-order approximation is used, it can be shown that the second-order approximation leads to $\pi(X_i;\beta_0, \beta_1) \approx \pi(\beta_0) + \gamma_1 X_i + \gamma_2 X_i^2$, where $\gamma_2 = \beta_1^2[1 - \pi(\beta_0)]^2[1 - 2\pi(\beta_0)]/2$. This illustrates that the quadratic linear regression equation requires estimation of three parameters, yet for logistic regression only two parameters are estimated. Similar results hold for the logistic – minimax regression model. Hence, when the effect of covariates is non-linear, it may be more efficient to fit the logistic model than the polynomial linear regression model. When might we expect the effect of a covariate to be non-linear? This could arise when only a subset of the ASPs are linked, and both the mean and variance of the covariate differ between ASPs that are linked vs. not linked [31]. In fact, Kong and Cox [32] motivate an exponential allele sharing model for this type of situation. Despite these theoretical considerations, the functional dependence of IBD probabilities on covariates is typically unknown. Therefore, whether linear or nonlinear regression models are used, it would be wise to use standard model-fitting diagnostics, such as plots of residuals vs. predicted values, to examine model assumptions.

A difference between linear regression vs. the nonlinear logistic models is that the regression coefficients for linear regression should be bounded so that the fitted π values are bound between 0 and 1, because π values are proportions. If X is based on a categorical variable (e.g. exposed vs. unexposed), the average π values in subgroups defined by X are typically much closer to 0.5 than they are to 0.0 or 1.0, and thus fitted values by linear regression will not exceed these bounds. If, however, X is based on a continuous variable, sib pairs with a large X value could yield a fitted π that falls out of range. Although a constraint could be applied in fitting the linear regression model, standard software does not offer this option, and this constraint can be difficult to impose when there are multiple covariates in the model. In contrast, the logistic models naturally provide this constraint without bounding the estimated regression coefficients. However, if one wishes to constrain the z_i values to fall within the genetically possible triangle, then constraints must be imposed on the logistic regression coefficients [18, 24]. Whether the z_i values should be constrained may be debatable, because if the effects of the covariates are mismodeled, or important covariates are not included in the regression model, then the z_i sharing probabilities are not expected to fall within the 'possible triangle' [24, 33, 34].

Another feature favoring a logistic type of model is that it naturally handles heteroscedasticity. An assumption of linear regression is homoscedasticity, which in our case means that the variance of π, conditional on the covariates, does not depend on the values of the covariates. However, π is a proportion, based on trinomial probabilities, with variance $Var(\pi) = z_2(1 - z_2) + z_1(1 - z_1)/4 - z_1z_2$. If the z_i values depend on covariates, then $Var(\pi)$ will also depend on covariates, and the logistic model makes use of this heteroscedasticity, yet linear regression ignores it, and in fact is penalized by the occurrence of heteroscedasticity.

Relative Pairs from Pedigrees

Other important issues that arise when considering multiple relative pairs created from pedigrees are how to account for statistical dependencies among relative pairs created from the same pedigree, how to construct statistical tests for different types of relative pairs (beyond sib pairs), and how to combine these statistical tests into a summary statistic. First consider a quantitative trait regression analysis with sibships having more than two sibs. In this case, the creation of all possible pairs of sibs will create pairs that overlap by either 0, 1, or 2 sibs. The program *SIBPAL* in the software S.A.G.E. [35] estimates the correlation between pairs of sib-pairs according to the amount of overlap, with correlation coefficients of r_0 and r_1 for when 0 or 1 sibs overlap between the pairs of sib-pairs (r_2 = 1 when 2 sibs overlap). These coefficients are used to account for the correlations between pairs of sib-pairs, and further robust regression is attained by use of a generalized estimating equation (GEE), which uses a robust 'sandwich' variance estimate to account for the dependencies among sib-pairs. Thomas et al. [36] also proposed use of a GEE method for analyzing linkage in sibships. Their method allows for both person-specific and pair-specific covariates (which do not necessarily depend on each other), and accounts for correlation among sib-pairs. They use a first-order GEE to estimate parameters for person-specific covariates that influence the trait mean. Then, these residuals are used in a second-order GEE to estimate parameters that influence the covariances among siblings; these parameters can represent linkage (via IBD probabilities) as well as gene-environment and/or gene-gene interactions. Extending either of these GEE approaches to different types of relative pairs created from a pedigree is conceptually straightforward, but would require additional parameters to account for the different types of relative pairs.

An alternative approach was taken by Sham et al. [5] (see also [4]) in order to simultaneously account for all possible types of relative pairs from a pedigree. They use multivariate regression, which is based on covariance matrices. They derive a matrix for covariances of the squared-trait differences and the squared-trait sums, based on the kinship matrix for a pedigree, the assumption of multivariate normality of the trait vector for the pedigree, as well as a specified trait heritability. This approach avoids the need to estimate separate correlations for each category of relative pairs, but at the price of assumed values of population parameters (trait mean, variance, and heritability); mis-specification of the population parameters can have a large influence on the power of this approach. Other covariances are needed, such as the covariance of the estimated IBD sharing among pairs of relatives, the covariance of the squared-trait difference with the IBD sharing, and the covariance of the squared-trait sum with the IBD sharing. These covariance matrices are combined in a multivariate regression, which implicitly weights the different types of relative-pairs by weights that are estimated from the data, as well as implicitly accounts for the correlations among the different relative-pairs within a pedigree. These methods are implemented in the software Merlin [37].

For affected relative-pair studies, it is straight-forward to adapt the mean sharing statistic for different types of relative pairs, because the numerator of the statistic is ($\hat{\pi} - \pi_0$), where π_0 is this null value for the type of relative-pair (e.g., π_0 has values of 1/2 for sib-pairs, and 1/8 for first-cousin-pairs). This statistic can be created for each category of relative pairs, with each statistic standardized by its standard error. Then, an optimal way to combine the standardized statitistics for different categories of relative-pairs is a linear combination with weights that depend on the degree of relationship as well as the sibling relative risk, λ (assuming a monogenic trait) [38]. Although λ may not be well known, reasonable choices for λ are given by Teng and Siegmund [38], who show that mis-specification of the λ used to create the weights does not dramatically decrease the efficiency of the combined statistical test. This approach could be implemented as a weighted regression with intercepts that vary according to the type of relationship for a pair, such as $E[\hat{\pi}_r] = \pi_{0r}$, where π_{0r} is the null value for a pair of relatives of type r. Under the null hypothesis of no linkage, the residuals would have mean zero, and so the mean of these residuals could be used to construct a test statistic. To account for dependencies among pairs, a robust 'sandwich' variance estimate could be used. However, this approach does not seem to be incorporated into existing software. Further work is needed to evaluate the statistical properties of this type of regression, such as the influence of the proposed weights and the sandwich variance estimator. In contrast to the mean-sharing statistic, the conditional logistic regression model [18, 22] naturally accounts for different types of affected relative-pairs, because each type of pair is parameterized using only the offspring recurrence risk ratio. Other model-free lod score models require estimation of different sets of parameters for different types of relative-pairs, making it more difficult to combine different types of relative-pairs. The influence of residual correlations among pairs of affected sib-pairs on the likelihood ratio statistics from these logistic models does not seem large, because it has been shown that multiple pairs within a sibship can be treated as independent pairs, provided the number of affected sibs in a sibship is not larger than six; some down-weighting is needed for larger sibships [39]. Whether this rule-of-thumb continues to hold when more than one type of relative pair and/or covariates are included in the model requires additional investigation.

Discussion

There have been many advances in the development of regression models for the analysis of linkage of traits, both quantitative and affection status, to genetic markers. One of the earliest regression methods, the HE regression, is in fact a variance component method, which can be extended to separate out additive and dominant genetic components by regressing on both $\hat{\pi}$ and $\hat{z}_2$. What is usually called a variance component method takes a likelihood approach based on the critical assumption of normality of the residuals (multivariate normality for pedigree data). In contrast, the regression approach offers appeal because of its robust properties. Furthermore, because the regression methods are rapidly computed, it is feasible to use a large number of permutations to compute empirical p values from the permutation distribution, which may be most important for the extremely small p values typically used to judge statistical significance for linkage studies. Nonetheless, one must be careful in the interpretation of the resulting parameter estimates and statistical inference, particularly when considering linkage heterogeneity, gene-environment interaction, or sampling ASPs without unaffected sib 'controls'. Sampling only ASPs requires that the genetic markers segregate in Mendelian fashion, and hence ASP analyses could be biased when markers are in linkage disequilibrium with genes that influence death prior to the age at which subjects are sampled. Furthermore, when considering the influence of environmental risk factors on linkage (as for evaluation of gene-environment interaction), it is necessary to assume that the gene and environmental risk factor are independent in the population. These points favor sampling some unaffected siblings to be sure that conclusions are not biased by invalid assumptions. Elston et al. [40] have suggested recruiting, when conducting an ASP study, an additional unaffected sibling for half the ASPs; this results in an equal number of ASPs and DSPs for an increase in sample size of only 25%.

We have emphasized that the definition of the trait function (e.g., $(y_1 - y_2)^2$, or the mean-corrected product, $y_1 \times y_2$) is critical for linkage analysis of quantitative traits. We also emphasized that lod scores do not change when interchanging Y and $\hat{\pi}$ in the linear regression equation, and that this 'trick' can be used to evaluate how best to include covariates in regression models for affected sib pairs. In particular, covariates defined as $(x_1 - x_2)$ or $(x_1 + x_2)$ measure 'pure' heterogeneity in linkage of affection status. That is, these types of covariates measure heterogeneity of IBD sharing that can be caused by linkage heterogeneity or gene-environment interaction, but they are not confounded by linkage of the x trait. In contrast, the covariate $(x_1 - x_2)^2$ can be confounded when x is linked.

The contrast of linear regression models with the recently developed non-linear logistic regression models for ASPs emphasized the strengths and weaknesses of both methods. The linear model was shown to be a natural extension of the commonly used mean test for including one or more covariates to test for heterogeneity in IBD sharing. When the distribution of covariates differs between subsets of ASPs that are linked and not linked, the nonlinear logit models may be most efficient at detecting the influence of the covariate on IBD sharing. Also, the logistic models are better than the linear model to handle heteroscedasticity. Further work is needed to compare the relative merits of the linear vs. nonlinear methods, such as their relative power and robustness, as well as the interpretation of regression coefficients.

We discussed several different approaches to account for multiple relative pairs from the same pedigree. For sibling pairs, the GEE methods are straightforward. However, more work is required to extend them to larger pedigrees, and evaluate their statistical properties, particularly when some types of relative pairs are infrequent. In contrast to GEE, the logistic regression methods for affected relative pairs directly model different types of relative pairs through the offspring recurrence risk ratio. Although this seems to offer promise, more work is required to evaluate its properties for larger extended pedigrees.

Overall, it is important to recognize that the regression-based methods provide flexibility in the number of covariates in the model, as well as how each covariate is modeled (e.g., linear, quadratic, etc.). However, mis-specification of the functional relationship between covariates and IBD sharing probabilities can weaken power. If the linkage signal is strongest within a subset described by a specific combination of covariate data, this subset could be missed if the appropriate interaction terms are not included in the regression model. Further work is needed to evaluate the best ways to model the effects of covariates, particularly whether the potential gain by having additional regression parameters is worth the price in terms of additional degrees of freedom, and potential loss in power.

Acknowledgements

This research was supported by United States Public Health Services, National Institutes of Health; Contract grant numbers DE13276, DK57292, ES10421, GM28356, GM65450, HG01577, and RR03655.

Appendix

t Statistic for Slope Is Invariant with Respect to Interchange of y and x

The t statistics for the slope b_{yx} for regression of y on x, for the slope b_{xy} for regression of x on y, and for Pearson's correlation coefficient $\hat{\rho}$, can all be expressed as

$$t_{N-2} = \frac{\sqrt{(N-2)}\, C_{xy}}{\sqrt{(V_x V_y - C_{xy}^2}}$$

where

$$C_{xy} = Cov(x,y) = \sum (x - \overline{x})(y - \overline{x})/N\,,$$

$$V_x = Var(x) = \sum (x - \overline{x})^2/N\,,$$

and

$$V_y = Var(y) = \sum (y - \overline{y})^2/N\,.$$

This illustrates that the t statistic is invariant with respect to regression of y on x vs. the regression of x on y. The relationship of these regression coefficients with Pearson's correlation coefficient can be summarized as

$$b_{yx} = C_{xy}/V_x\,,$$

$$b_{xy} = C_{xy}/V_y\,,$$

$$\hat{\rho}^2 = \frac{C_{xy}^2}{V_x V_y} = \frac{b_{yx}^2\, V_x}{V_y} = \frac{b_{xy}^2\, V_y}{V_x}\,.$$

Although the t statistic for the slope is invariant, the regression parameters and their variances of course depend on whether y is regressed on x, or x regressed on y.

References

1 Haseman JK, Elston RC: The investigation of linkage between a quantitative trait and a marker locus. Behav Genet 1972;2:3–19.
2 Zhang W, Collins A, Lonjou C, Morton N: A linkage tournament: affection status, parametric analysis, multivariate traits, and enhancements to variance components and relative pairs. Ann Hum Genet 2002;66:87–98.
3 Visscher P, Hopper J: Power of regression and maximum likelihood methods to map QTL from sib-pair and DZ twin data. Ann Hum Genet 2001;65:583–601.
4 Putter H, Sandkuijl L, van Houwelingen J: Score test for detecting linkage to quantitative traits. Genet Epidemiol 2002;22:345–355.
5 Sham P, Purcell S, Cherny S, Abecasis G: Powerful regression-based quantitative-trait linkage analysis of general pedigrees. Am J Hum Genet 2002;71:238–253.
6 Feingold E: Regression-Based Quantitative-Trait Locus Mapping in the 21st Century. Am J Hum Genet 2002;71:217–222.
7 Drigalenko E: Letter to the Editor: How sib pairs reveal linkage. Am J Hum Genet 1998; 63:1242–1245.
8 Elston RC, Buxbaum S, Jacobs KB, Olson JM: Haseman and Elston revisited. Genet Epidemiol 2000;19:1–17.
9 Xu X, Weiss S, Xu X, Wei LJ: A unified Haseman-Elston method for testing linkage with quantitative traits. Am J Hum Genet 2000;67: 1025–1028.
10 Forrest W: Weighting improves the 'New Haseman-Elston' method. Hum Hered 2001;52:47–54.
11 Sham PC, Purcell S: Equivalence between Haseman-Elston and Variance-Components Linkage Analyses for Sib Pairs. Am J Hum Genet 2001;68:6.
12 Feingold E: Methods for linkage analysis of quantitative trait loci in humans. Theoret Pop Biol 2001;60:167–180.
13 Palmer L, Jacobs K, Elston R: Letter to the Editor – Haseman and Elston revisited: The effects of ascertainment and residual familial correlations on power to detect linkage. Genet Epidemiol 2000;19:456–460.
14 Shete S, Chen J, Zhou X, Amos C: Modeling age x major gene interaction by a variance component approach. Genet Epidemiol 2001;21: S849-S853.
15 Gauderman W, Morrison J, Siegmund K: Should we consider gene x environment interaction in the hunt for quantitative trait loci? Genet Epidemiol 2001;21:S831-S836.
16 Blackwelder WC, Elston RC: A comparison of sib-pair linkage tests for disease susceptibility loci. Genet Epidemiol 1985;2:85–97.
17 Gauderman W, Siegmund K: Gene-environment interaction and affected sib pair linkage analysis. Hum Hered 2000;52:34–46.
18 Goddard KA, Witte JS, Suarez BK, Catalona WJ, Olson JM: Model-free linkage analysis with covariates confirms linkage of prostate cancer to chromosomes 1 and 4. Am J Hum Genet 2001;68:1197–206.
19 Witte J, Goddard K, Conti D, Elston R, Lin J, Suarez B, Broman K, Burmester J, Weber J, Catalona W: Genomewide scan for prostate cancer-aggressiveness loci. Am J Hum Genet 2000;67:92–99.

20 Khoury MJ, Flanders WD: Nontraditional epidemiologic approaches in the analysis of gene-environment interactions: Case-control studies with no controls! Am J Epidemiol 1996;144: 207–213.
21 Risch N: Linkage strategies for genetically complex traits. III The effect of marker polymorphism on analysis of affected relative pairs. Am J Hum Genet 1990;46:242–253.
22 Olson JM: A general conditional-logistic model for affected-relative-pair linkage. Am J Hum Genet 1999;65:1760–1769.
23 Greenwood CMT, Bull SB: Incorporation of covariates into genome scanning using sib-pair analysis in bipolar affective disorder. Genet Epidemiol 1997;14:635–640.
24 Greenwood CMT, Bull SB: Analysis of affected sib pairs, with covariates – with and without constraints. Am J Hum Genet 1999;64:871–885.
25 Bull S, Greenwood C, Mirea L, Morgan K: Regression models for allele sharing: analysis of accumulating data in affected sib pair studies. Stat Med 2002;21:431–444.
26 Rice J: Commentary: The role of meta-analysis in linkage studies of complex traits. Am J Med Genet 1997;74:112–114.
27 Morton NE: Logarithm of odds (lods) for linkage in complex inheritance. Proc Natl Acad Sci USA 1996;93:3471–3476.
28 Rice J, Rochberg N, Neuman R, Saccone N, Liu K-Y, Zhang X, Culverhouse R: Covariates in linkage analysis. Genet Epidemiol 1999;17: 691–695.
29 Whittemore AS, Tu I-P: Simple, robust linkage tests for affected sib pairs. Am J Hum Genet 1998;62:1228–1242.
30 Holmans P: Asymptotic properties of affected-sib-pair linkage analysis. Am J Hum Genet 1993;52:362–374.
31 O'Brien P: Comparing two samples: Extensions of the *t*, rank-sum, and log-rank test. J Am Statist Assoc 1988;83:52–61.
32 Kong A, Cox NJ: Allele-sharing models: LOD scores and accurate linkage tests. Am J Hum Genet 1997;61:1179–1188.
33 Guo S-W: Gene-environment interactions and the affected-sib-pair designs. Hum Hered 2000;50:271–285.
34 Dizier M, Quesneville H, Prum B, Selinger-Leneman H, Clerget-Darpoux F: The triangle test statistic (TTS): a test of genetic homogeneity using departure from the triangle constraints in IBD distribution among affected sib-pairs. Ann Hum Genet 2000;64:433–442.
35 S.A.G.E., 'Statistical Analysis for Genetic Epidemiology,' Cork, Statistical Solutions Ltd., 2002.
36 Thomas D, Qian D, Gauderman W, Siegmund K, Morrison J: A generalized estimating equations approach to linkage analysis in sibships in relation to multiple markers and exposure factors. Genet Epidemiol 1999;17:S737-S742.
37 Abecasis G, Cherny S, Cookson W, Cardon L: Merlin – rapid analysis of dense genetic maps using sparse gene flow trees. Nat Genet 2002; 30:97–101.
38 Teng J, Siegmund D: Combining information wi12,8 hin and between pedigrees for mapping complex traits. Am J Hum Genet 1997;60: 979–992.
39 Greenwood C, Bull S: Down-weighting of multiple affected sib pairs leads to biased likelihood-ratio tests, under the assumption of no linkage. Am J Hum Genet 1999;64:1248–1252.
40 Elston RC, Guo X, Williams LV: Two-stage global search designs for linkage analysis using pairs of affected relatives. Genet Epidemiol 1996;13:535–558.

Human Heredity

Hum Hered 2003;55:97–107
DOI: 10.1159/000072314

X-Linked Extension of the Revised Haseman-Elston Algorithm for Linkage Analysis in Sib Pairs

Howard Wiener[a] Robert C. Elston[b] Hemant K. Tiwari[c]

[a]Department of Epidemiology, University of Alabama at Birmingham, Birmingham, Ala., [b]Department of Epidemiology and Biostatistics, Rammelkamp Center for Education and Research, Case Western Reserve University, Cleveland, Ohio, [c]Section on Statistical Genetics, Department of Biostatistics and Molecular and Cellular Pathology, Department of Pathology, University of Alabama at Birmingham, Birmingham, Ala., USA

Key Words

X-linkage · Sib pairs · Haseman-Elston · Model-free · IBD · QTL

Abstract

Haseman and Elston (H-E) proposed a regression-based robust test of linkage between a marker and an autosomal quantitative trait locus, using the squared sib pair trait difference as a dependent variable and the proportion of alleles shared identical by descent by the sib pair as an independent variable. Several authors have proposed improvement of the original H-E's seminal work by using an optimal linear combination of squared sum and squared difference as the dependent variable. In this paper, we extend Haseman and Elston's sib pair method to an X-linked locus. We give a general formulation of the complete regression model and details of the regression coefficients in terms of variance components. Simulation results are presented to describe the power of this technique for a theoretical best case scenario.

Introduction

Haseman and Elston [1972] introduced a sib pair method using classical regression analysis to detect linkage between a polymorphic marker locus and any quantitative trait locus. In particular, Haseman and Elston [1972] proposed regressing the squared sib pair trait difference on the proportion of alleles the sibs share identical by descent (i.b.d.) at a marker locus, this proportion being estimated from the sibs' and their parents' marker phenotypes. Wright's [1997] and Drigalenko's [1998] letters to the editor in the American Journal of Human Genetics, suggested using both squared sib pair trait differences and squared sib pair trait sums to capture all the information and so increase the power of the test. This led to an avalanche of papers, suggesting extensions to the test by using different weighting schemes to combine squared sums and squared differences to achieve optimal power for autosomal loci [see Shete et al.,1999; Elston et al., 2000; Xu et al., 2000; Forrest, 2001; Sham and Purcell, 2001; Sham et al., 2002; Furest, Feingold; Visscher and Hopper, 2001, to name a few]. But, the theoretical development of the Haseman-Elston extension to X-linked loci has not been described in the statistical genetics literature. How-

KARGER

Fax + 41 61 306 12 34
E-Mail karger@karger.ch
www.karger.com

0001–5652/03/0553–0097$19.50/0

Accessible online at:
www.karger.com/hhe

Dr. Howard Wiener
Department of Epidemiology
RYALS-217D
Birmingham, AL 35294-0022 (USA)
Tel. +1 205 934 9088, Fax +1 205 975 4520, E-Mail HWiener@uab.edu

ever, the software packages for model-free linkage analyses like Allegro [Gudbjartsson et al., 2000]; GENEHUNTER [Kruglyak et al., 1996]; MINX, an X chromosome specific version of Merlin [Abecasis, 2002]; SOLAR [Blangero and Almasy, 1996; Almasy and Blangero, 1998] – have implemented X-chromosome statistics, without giving any information on the algorithms used in the software. Here is the quote from Merlin documentation: 'MINX (MERLIN in X) is an X-specific version of Merlin. It is available in distributions of MERLIN version 0.9.1 and later. There is currently no manuscript describing MINX performance and algorithms in detail. Although I believe MINX results to be correct, the methods are unpublished and I would advise using with care'. Also, GENEHUNTER has dropped the X-linkage option after version 1.3. In view of the paucity of information available for the X-linked version of the original H-E, as well as its extensions, we give a general formulation of the complete regression model and details of the regression coefficient for sib pair data.

The purpose of this paper is to derive in detail all the necessary model components for a sex-linked analog of the various forms of the Haseman and Elston [1972] and Elston et al. [2000] regression linkage tests. In order to model the linkage between a trait and a putative genetic locus, we must derive the correlation between the inheritance at that locus and the expression of the trait. For linkage analysis, we typically do not have a priori information about the relationship between a functional locus and the given trait, and we must depend on information from a marker. When the genetic distance between the marker and the functional gene affecting the expression of the trait is small, the correlation between the trait and the functional gene will be mirrored by a correlation between the trait and inheritance of the marker. Pairs of siblings can be compared to calculate this correlation. We assume that there is some numerical measure of the trait expression, which may reduce to two values for an affection status, but need not. One quantity that can be used to express trait inheritance in the analysis is the square of the difference of the trait values of the two siblings in the pair, as originally discussed in Haseman and Elston [1972]. There is additional information available in the square of the mean corrected sum of the traits of the sibs, and in linear combinations of these two squares, as discussed in Shete et al. [1999] and Elston et al. [2000]. The inheritance of the marker will be expressed in terms of identity by descent (IBD): the number of alleles at the given locus that the two siblings inherited from the same parental chromosome. The symbol Ω is used here to denote IBD. We will proceed in 4 steps to derive an expression for this correlation. First, we will derive expressions for the variance of the trait expression; later results will be expressed in terms of these quantities. Second, we will derive the relationship between inheritance at the putative functional locus and the expression of the trait. The third step will be to derive correlations between the unobservable inheritance of the functional locus and the observed inheritance at a marker. This will be conditional on the recombination frequency between the unknown functional locus and the observed marker locus. This will be used to derive the relationship between trait expression and marker inheritance. Finally, we will derive the IBD of a sibling pair based on their observed genotypes.

Methods

The relationship between trait expression in two siblings and the IBD status at a marker locus between these siblings can be expressed in terms of a regression model. The value of the regression parameters will depend on the recombination fraction between the unobserved functional locus and the observed marker locus. This is the basis of the technique to explore the linkage between the trait and the marker.

Part I: Calculation of Trait Variance

Let i index the sib pair, and $j = 1, 2$ indicate the individual sib. Let the trait locus have 2 alleles, B and b, with frequencies p and $q = 1 - p$, respectively. For individual j in sib pair i, let the mean value for the trait be $\mu + g_{ij}$ where g_{ij} is the component of the trait value due to the genotype at this locus. The complete trait value is given by $x_{ij} = \mu + g_{ij} + e_{ij}$ where e_{ij} is the component due to all other causes: genetic background, environmental interactions, and chance. Women have three possible genotypes: bb, Bb, and BB. The three values of g_{ij} corresponding to these genotypes will be denoted $-a$, d, and a, respectively. Note that this subsumes dominant, recessive, and additive models. In particular, $d = -a$ implies that B is Mendelian recessive, $d = a$ implies that B is Mendelian dominant, and $d = 0$ implies an additive effect of B. Assuming Hardy-Weinberg equilibrium for the frequencies of the possible genotypes at this locus, the frequencies of bb, Bb, and BB are q^2, $2pq$, and p^2, respectively. Given these frequencies, $E(G) = (-a)q^2 + (d)2pq + (a)p^2 = a(p^2 - q^2) + 2pqd = a(p - q) + 2pqd$. In the same way, we find that $E(G^2) = a^2(p^2 + q^2) + 2pqd^2$, from which it follows that the trait variance is given by $V(G) = 2pq[a - d(p - q)]^2 + 4p^2q^2d^2$. We denote the two terms in this sum σ_a^2 and σ_d^2, respectively.

In order to derive the corresponding values for a male, note that there will only be two possible genotypes: B and b. The corresponding values of g are taken to be a and $-a$, respectively. Since there are only two possible genotypes, there is no distinction between an additive model, a dominant model, and a recessive model: two points always determine a line. As above, we have for males: $E(G) = a(p - q)$, $E(G^2) = a^2$ and finally, $V(G) = 4pqa^2$. We will denote this value σ_m^2. If the model for a female is entirely additive (i.e. if $d = 0$), the formula for the variance reduces to $2pqa^2$, exactly half the value noted for males.

Part II: Conditional Expectation of the Squared Trait Difference (Y) and the Squared Mean-Corrected Trait Sum (Z) on IBD (Ω) for a Functional Locus

Recalling that the observed value of the trait is x_{ij} for individual j in sibship i (j = 1, 2), and setting $d_{ij} = x_{ij} - \bar{x}$, we can use any of several functions to indicate the relationship between the phenotypes of two siblings. We can assume that $\bar{x}$ has a mean equal to μ, so the effect of this subtraction is to remove the μ term from the computations at the expense of complicating the error term (e_{ij}). The various generalized forms of the Haseman-Elston technique all use linear combinations of two variables. One variable of interest is given by $y_i = (d_{i1} - d_{i2})^2$ and the other is given by $z_i = (d_{i1} + d_{i2})^2$.

When we are dealing with the genetics of a sex-linked locus for a pair of siblings, we must make separate calculations for pairs of sisters, pairs of brothers, and brother-sister pairs. Consider first the case of two sisters. They necessarily inherited the same X chromosome from their father, so they necessarily have at least one allele identical by descent at an X chromosome locus. Therefore, the IBD value (Ω) at a locus between two sisters is necessarily either 1 or 2. If Ω = 2, the probability of the given sibling pair occurring is the probability of one pair of alleles occurring at random. For example, the probability of observing a pair of siblings, both with genotypes Bb, that are completely identical by descent (i.e. with Ω = 2) is the probability of choosing at random the pair of alleles B and b. This probability is $2pq$. If Ω = 1, the probability of observing the sibling pair is the probability of choosing three alleles at random. For example, the probability of observing a pair of siblings with genotypes BB and Bb with Ω = 1 is the probability of observing an individual of type BB, and then choosing another single allele of type b. This probability is p^2q. Table 1 summarizes the values of the squared trait difference and sum for each possible sibling type, and their corresponding probabilities. Since Ω = 0 is impossible for two full sisters, sister pairs with types BB/bb are impossible. In this table, the stochastic terms are defined as $e_i = e_{i1} - e_{i2}, f_i = e'_{i1} + e'_{i2}$, where the primes indicate that the error term for estimation of μ by $\bar{x}$ has been included. Using the values summarized in this table, we can calculate $E(Y|\Omega)$. In particular, we note that

$$E(Y|\Omega = 2) = \sigma_e^2(p^2 + 2pq + q^2) = \sigma_e^2.$$

Similarly,

$$E(Y|\Omega = 1) = \sigma_e^2 + \sigma_a^2 + 2\sigma_d^2.$$

We can also calculate the conditional expectations of Z on the two values of Ω.

$$E(Z|\Omega = 2) = a^2[4(p^2 + q^2)] + d^2(8pq) + \sigma_f^2,$$

and

$$E(Z|\Omega = 1) = a^2(4p^2\ 2pq + 4q^2) + d^2(6pq) + ad[4pq(p - q)] + \sigma_f^2.$$

Since males have only one allele, a pair of brothers can only have zero or one allele IBD. Table 2 gives the information for brother pairs. Based on the contents of this table, we find that

$$E(Y|\Omega = 1) = \sigma_e^2,$$

and similarly,

$$E(Y|\Omega = 0) = \sigma_e^2 + 2\sigma_m^2,$$

$$E(Z|\Omega = 1) = 4a^2 + \sigma_f^2,$$

Table 1. Outcome variables for sister pairs, and the frequencies of each genotype pair conditional on sharing IBD

Sister pair	Y	Z	Ω = 2	Ω = 1
BB-BB	e_i^2	$(2a + f_i)^2$	p^2	p^3
BB-Bb	$(a - d + e_i)^2$	$(a + d + f_i)^2$	0	p^2q
Bb-BB	$(d - a + e_i)^2$	$(a + d + f_i)^2$	0	p^2q
Bb-Bb	e_i^2	$(2d + f_i)^2$	$2pq$	$p^2q + pq^2$
Bb-bb	$(a + d + e_i)^2$	$(d - a + f_i)^2$	0	pq^2
bb-Bb	$(-a - d + e_i)^2$	$(d - a + f_i)^2$	0	pq^2
bb-bb	e_i^2	$(-2a + f_i)^2$	q^2	q^3

Table 2. Outcome variables for brother pairs, and the frequencies of each genotype pair conditional on sharing IBD

Brother pair	Y	Z	Ω = 1	Ω = 0
B-B	e_i^2	$(2a + f_i)^2$	p	p^2
B-b	$(2a + e_i)^2$	f_i^2	0	pq
b-B	$(-2a + e_i)^2$	f_i^2	0	pq
b-b	e_i^2	$(-2a + f_i)^2$	q	q^2

Table 3. Outcome variables for brother-sister pairs, and the frequencies of each genotype pair conditional on sharing IBD

Sister-brother pair	Y	Z	Ω = 1	Ω = 0
BB-B	e_i^2	$(2a + f_i)^2$	p^2	p^3
BB-b	$(2a + e_i)^2$	f_i^2	0	p^2q
Bb-B	$(d - a + e_i)^2$	$(a + d + f_i)^2$	pq	$2p^2q$
Bb-b	$(d + a + e_i)^2$	$(d - a + f_i)^2$	pq	$2pq^2$
bb-B	$(-2a + e_i)^2$	f_i^2	0	pq^2
bb-b	e_i^2	$(-2a + f_i)^2$	q^2	q^3

and

$$E(Z|\Omega = 0) = 4a^2 - 8pqa^2 + \sigma_f^2.$$

When we consider a brother-sister pair, we note that since the brother only has one allele, IBD cannot exceed 1. Table 3 covers the case for brother-sister pairs. Following the procedure used for other sibling pair types, we find that

$$E(Y|\Omega = 1) = \sigma_e^2 + 2pq(a^2 + d^2)$$
$$= \sigma_e^2 + \sigma_a^2 + 2\sigma_d^2 + 4pqad(p - q).$$

Similarly,

$$E(Y|\Omega = 0) = \sigma_e^2 + 2pq[a - d(p - q)]^2 + 8p^2q^2d^2 + 4pqa^2$$
$$= \sigma_e^2 + \sigma_a^2 + 2\sigma_d^2 + \sigma_m^2,$$

$$E(Z|\Omega = 1) = a^2(4p^2 + 2pq + 4q^2) + 2pqd^2 + \sigma_f^2,$$

Table 4. Mating types, and their population frequencies

I	ii × i	p_i^3
II	ii × j	$p_i^2 p_j$
III	ij × i	$2 p_i^2 p_j$
IV	ij × k	$2 p_i p_j p_k$

Table 5. Sibling pair types, and their probabilities conditional on mating types

	I	II	III	IV
1	1/4	0	1/16	0
2	0	0	1/8	0
3	0	1/4	1/16	1/8
4	0	0	0	1/8
5	1/4	1/4	1/8	1/8
6	0	0	1/8	1/8
7	1/2	0	1/8	0
8	0	0	1/8	0
9	0	1/2	1/4	1/4
10	0	0	0	1/4

Table 6. IBD sharing probabilities by parental mating type and sibling pair type

Parents	Siblings	f_0	f_1	f_2
I	1	0	1/2	1/2
	5	1/2	1/2	0
	7	1/2	1/2	0
II	3	0	1/2	1/2
	5	1/2	1/2	0
	9	1/2	1/2	0
III	1	0	0	1
	2	0	1	0
	3	0	0	1
	5	0	1	0
	6	1	0	0
	7	0	1	0
	8	1	0	0
	9	1	0	0
IV	3	0	0	1
	4	0	1	0
	5	0	1	0
	6	1	0	0
	9	0	1	0
	10	1	0	0

and

$$E(Z \mid \Omega = 0) =$$
$$a^2(4p^2 - 2pq + 4q^2) + d^2(2pq) + ad[4pq(p-q)] + \sigma_f^2.$$

As we might expect, the expected value of Y, which is essentially a measure of discordance between the measured trait values of the siblings in the pair, increases with a decrease of sharing IBD. This is clear for the same gender pairs. For the brother-sister pairs, we note that the difference $\sigma_m^2 - 4pqad(p-q)$ is equal to $4pqa[a - d(p-q)]$. We note that the term in brackets is necessarily between $(a-d)$ and $(a+d)$. If we only consider models for which $|d| \leq a$ – which include the additive, dominant, and recessive models as special cases, and excludes the case of heterosis – this is true for the brother-sister case as well.

Part III: Estimation of the IBD at the Marker Locus

In order to form the estimate of IBD sharing we note that there are only four mating types possible with respect to an X-linked locus. The four types and their frequencies are given in table 4. There are four possible types of pairs of sisters (since they must share one allele at the locus in question), two possible types of pairs of brothers, and four possible types of brother/sister pairs. These are given by: (1) ii-ii; (2) ii-ij; (3) ij-ij; (4) ij-ik; (5) i-i; (6) i-j; (7) ii-i; (8) ii-j; (9) ij-i, and (10) ij-k. The probabilities of the occurrence of each of the ten sibling pair types conditional on the parental mating type is given in table 5. It is interesting that there are four sibling pair types (2, 4, 8, and 10) that unambiguously define the parental mating type, and the precise parental genotypes are known for three of these four. This table will be used to help calculate IBD sharing when parental information is not available. When the parental genotypes are known, the calculation is straightforward. Consider a type I parental mating with a type 1 pair of daughters. We know that one of the '*i*' alleles in each of the sisters must have been the paternal '*i*' allele, so that Ω must be at least one. The two maternal '*i*' alleles are not distinguishable, so the maternal '*i*' allele in each daughter could have been either of the '*i*' alleles in the mother with equal probability. If we refer to these two alleles in the mother as 1 and 2, we can see that the maternal alleles in the two sisters could have been 1&1, 1&2, 2&1, or 2&2 with equal probability. Thus, the maternal '*i*' could be identical by descent or not with equal probability. If we use the notation f_1 for the probability that $\Omega = 1$ and f_2 for the probability that $\Omega = 2$, we see that in the case of type I parents with type 1 daughters, $f_1 = f_2 = 1/2$. The result of applying this reasoning to all of the parent/child quartets is given in table 6.

The calculation of IBD sharing between sibs in a sibship can be reduced to finding a weighted average of values looked up from a table. The discussion of this process is given in the appendix.

Part IV: Relationship between IBD at the Trait (Ω_t) and Marker (Ω_m) Loci, and the Conditional Expectation of Y and Z on IBD at the Marker Locus

We must remember that the reason for doing a linkage analysis is to determine whether there is evidence for the existence of a trait locus at a particular position in the genome, so that there cannot be any direct observation of the genotype of the (putative) trait locus. The information available for each individual in the study is the set of apparent genotypes at chosen marker loci. We assume linkage equilibrium between the marker locus under discussion and the putative trait locus, as was done in the autosomal case by Haseman and Elston. This assumption insures that the IBD value at the putative trait locus is a function of the IBD value at the marker locus and the distance between the marker and trait loci expressed as a recombination probability.

Consider first the case of a pair of sisters. Since they must have inherited the same X chromosome from their father, $\Omega \geq 1$ for both the trait and marker loci. The entire paternal X chromosome is assumed to be inherited unchanged. (This is not true of the pseudoautosomal region of the X chromosome, but our analysis is assuming that we are dealing only with true X-linked loci.) For this reason, we need only examine the maternal inheritance to determine the relationship between the IBD value at the marker locus (denoted Ω_m) and the IBD value at the trait locus (denoted Ω_t). If the mother's two-locus genotype is denoted A_1B_1/A_2B_2 (where the 'A' alleles are the marker alleles, and the 'B' alleles are the trait alleles), and c denotes the recombination fraction between the two loci, then the proportion of female gametes with genotypes A_1B_1, A_2B_2, A_1B_2, and A_2B_1 are respectively $(1-c)/2$, $(1-c)/2$, $c/2$ and $c/2$. In order for two sisters to have $\Omega_m = 2$ and $\Omega_t = 2$ (recalling that the paternal X chromosome is identical), their genotypes on the maternal X chromosome could be (A_1B_1,A_1B_1), (A_1B_2,A_1B_2), (A_2B_1,A_2B_1), or (A_2B_2,A_2B_2). The probabilities of these combinations are respectively $[(1-c)/2]^2$, $[c/2]^2$, $[c/2]^2$ and $[(1-c)/2]^2$. Thus $P(\Omega_m = 2 \text{ and } \Omega_t = 2)$ is the sum of these four terms, or $(1-c)^2/2 + c^2/2$, which we can write (following the notation of Haseman and Elston) as $\Psi/2$. We can summarize the enumeration of the pairs of genotypes that result by saying that the maternal X chromosomes in two sisters are IBD at two loci if there were the same number of recombinations between the loci (i.e. either 0 or 1), and the same chromosome was chosen for both gametes. If we had the same number of recombinations, but the opposite chromosome was chosen, we would have $\Omega_m = 1$ and $\Omega_t = 1$. It is straightforward to see that this would have resulted in the same probability: $\Psi/2$. The pairs that result in $\Omega_m = 2$ and $\Omega_t = 1$ are (A_1B_1,A_1B_2), (A_1B_2,A_1B_1), (A_2B_1,A_2B_2), and (A_2B_2,A_2B_1). In each case here, the number of recombinations that gave rise to the female gametes was different for the two sisters. The probability of each such pairing is $c(1-c)/4$. The case of $\Omega_m = 1$ and $\Omega_t = 2$ is symmetric to this case, so we see that $P(\Omega_m = 2 \text{ and } \Omega_t = 1) = P(\Omega_m = 1 \text{ and } \Omega_t = 2) = c(1-c)$. Table 7 summarizes the joint distribution of Ω_m and Ω_t for two sisters.

When we consider the case of two brothers, we note that they must receive the Y chromosome from their father, so there are only maternal X chromosomes to consider. The table that describes the joint distribution of Ω_m and Ω_t for that case can be derived from the above table by making one small change: in place of $\Omega_m = 1$, write $\Omega_m = 0$; in place of $\Omega_m = 2$, write $\Omega_m = 1$; do the same for Ω_t. The joint distribution of Ω_m and Ω_t for two brothers is then as given in table 8. We are finally ready to consider the case of a brother and a sister. The brother cannot inherit the paternal X chromosome, so we are again considering only the inheritance of the maternal X chromosomes. The joint distribution in this case is identical to the case of two brothers.

We are interested here in the expected value of the squared trait difference (Y) or squared mean corrected trait sum (Z) conditional on the value of Ω_m. The expectation of squared trait difference conditional on Ω_t was derived in the previous section, and the joint distribution of Ω_t and Ω_m is given above. Noting that, for every possible value of Ω_m, the probability of that value is 1/2, it is simple to see that the conditional probability $P(\Omega_t = a \mid \Omega_m = b)$ is just twice the joint probability $P(\Omega_t = a \cap \Omega_m = b)$. Now, for the case of two sisters, we have $E(Y \mid \Omega_m = 1) = E(Y \mid \Omega_t = 1)P(\Omega_t = 1 \mid \Omega_m = 1) + E(Y \mid \Omega_t = 2)P(\Omega_t = 2 \mid \Omega_m = 1)$. Using the results from the previous section and contents of the above tables, we have $E(Y \mid \Omega_m = 1) = (\sigma_e^2 + \sigma_a^2 + 2\sigma_d^2)\Psi + (\sigma_e^2)2c(1-c)$. This simplifies slightly to $\sigma_e^2 + (\sigma_a^2 + 2\sigma_d^2)\Psi$. The reasoning is similar for the other sibling pair types and other possible values of Ω_m as well as for the other outcome variable (Z). We summarize all the results for Y in table 9. The results for Z are in table 10.

These relationships can be subsumed into a single model by using two variables: a classification variable and a continuous variable. The classification variable will have three levels, and will be included in a model either by using two dummy variables, or by allowing the model to be less than full rank. The variable, which we will denote as h, will have the value 4 for a pair of sisters, 2 for a pair of brothers, and 3 for a sister/brother pair. This choice of values corresponds to h being the sum of the gender codes for the siblings (male = 1, female = 2). The continuous variable, which we will denote as ω, will be defined in the following way. Let $\omega = 0$ at a locus for a sibling pair if the amount of sharing IBD at that locus is the maximum possible value for that sibling pair (i.e. 2 for a pair of sisters, 1 for the other gender combinations). A positive value of ω indicates a decrease of

Table 7. Joint distribution of IBD sharing at marker and trait loci for a pair of sisters

	$\Omega_m = 1$	$\Omega_m = 2$	Total
$\Omega_t = 1$	$\Psi/2$	$c(1-c)$	1/2
$\Omega_t = 2$	$c(1-c)$	$\Psi/2$	1/2
Total	1/2	1/2	

Table 8. Joint distribution of IBD sharing at marker and trait loci for a pair of brothers or a brother-sister pair

	$\Omega_m = 0$	$\Omega_m = 1$
$\Omega_t = 0$	$\Psi/2$	$c(1-c)$
$\Omega_t = 1$	$c(1-c)$	$\Psi/2$

Table 9. Expected value of squared difference outcome conditional on sibling pair type and IBD sharing at marker locus

Genders	Ω_m	$E(Y \mid \Omega_m)$
F/F	1	$\sigma_e^2 + (\sigma_a^2 + 2\sigma_d^2)\Psi$
F/F	2	$\sigma_e^2 + (\sigma_a^2 + 2\sigma_d^2)2c(1-c)$
M/M	0	$\sigma_e^2 + 2\sigma_m^2\Psi$
M/M	1	$\sigma_e^2 + 4\sigma_m^2 c(1-c)$
M/F	0	$(\sigma_e^2 + \sigma_a^2 + 2\sigma_d^2) + \sigma_m^2\Psi + 8pqad(p-q)c(1-c)$
M/F	1	$(\sigma_e^2 + \sigma_a^2 + 2\sigma_d^2) + 2\sigma_m^2 c(1-c) + 4pqad(p-q)\Psi$

Table 10. Expected value of squared mean-corrected sum outcome conditional on sibling pair type and IBD sharing at marker locus

Genders	Ω_m	$E(Z\|\Omega_m)$
F/F	1	$\sigma_f^2 + (1/pq)(\sigma_m^2 + 2\sigma_d^2) - (2\sigma_a^2 + 4\sigma_d^2 + 3\sigma_m^2)\Psi$
F/F	2	$\sigma_f^2 + (1/pq)(\sigma_m^2 + 2\sigma_d^2) - (2\sigma_a^2 + 4\sigma_d^2 + 3\sigma_m^2)2c(1-c)$
M/M	0	$\sigma_f^2 + (1/pq)\sigma_m^2 - 4\sigma_m^2\Psi$
M/M	1	$\sigma_f^2 + (1/pq)\sigma_m^2 - 8\sigma_m^2 c(1-c)$
M/F	0	$\sigma_f^2 + \sigma_m^2/(pq) - (2c-1)^2\sigma_m^2/2 + \sigma_d^2/(2pq) + 4pqad(p-q)\Psi$
M/F	1	$\sigma_f^2 + \sigma_m^2/(pq) + (2c-1)^2\sigma_m^2/2 + \sigma_d^2/(2pq) + 8pqad(p-q)c(1-c)$

Table 11. Expected value of arbitrary linear combination of Y and Z outcome variables conditional on sibling pair type and IBD sharing at marker locus

Genders	Ω_m	$E(C\|\Omega_m)$
F/F	1	$\sigma_c^2 - (s/pq)(\sigma_m^2 + 2\sigma_d^2) + ((r + 2s)(\sigma_a^2 + 2\sigma_d^2) + 3s\sigma_m^2)\Psi$
F/F	2	$\sigma_c^2 - (s/pq)(\sigma_m^2 + 2\sigma_d^2) + ((r - 2s)(\sigma_a^2 + 2\sigma_d^2) - 3s\sigma_m^2)2c(1-c)$
M/M	0	$\sigma_c^2 - (s/pq)\sigma_m^2 + (1/pq)\sigma_m^2 + (2r + 4s)\sigma_m^2\Psi$
M/M	1	$\sigma_c^2 - (s/pq)\sigma_m^2 + (1/pq)\sigma_m^2 + (2r + 4s)\sigma_m^2 2c(1-c)$
M/F	0	$\sigma_c^2 - (s/pq)(\sigma_m^2 + \sigma_d^2/2) + r(\sigma_a^2 + 2\sigma_d^2) + s(2c-1)^2\sigma_m^2/2$ $+ rpqad(p-q)2c(1-c) + (r\sigma_m^2 - 4spqad(p-q))\Psi$
M/F	1	$\sigma_c^2 - (s/pq)(\sigma_m^2 + \sigma_d^2/2) + r(\sigma_a^2 + 2\sigma_d^2) - s(2c-1)^2\sigma_m^2/2$ $+ rpqad(p-q)\Psi + (r\sigma_m^2 - 4spqad(p-q))2c(1-c)$

IBD sharing below the maximum value, so that $\omega = 1$ for the minimum value of IBD at the locus. The true value of ω can only be 0 or 1, but the estimate will be able to take any value between these extremes. The model that is appropriate for this situation is:

$$Y = \mu + \alpha_1 I_1(h) + \alpha_2 I_2(h) + \alpha_3 I_3(h) + \beta_1 I_1(h)\omega + \beta_2 I_2(h)\omega + \beta_3 I_3(h)\omega\,.$$

In this notation, $I_j(h)$ is the indicator function that has the value 1 when $h = j$, and 0 when $h \neq j$. In SAS notation, e.g. in Proc GLM or Proc Mixed, we can denote the model as $Y = h\omega\ (h)$. The test of interest would be a 3 d.f. test, i.e. a test of whether all three slopes in this linear model are significantly greater than zero. Note that the slopes are only meaningfully different from zero if they are positive: i.e. when an increased value of ω (i.e. a decrease in IBD sharing) mirrors an increase in sibling pair discordance.

In order to express the final result of applying the revisited Haseman-Elston method to this situation, it suffices to consider arbitrary linear combinations of the Y and Z outcome variables. To this end, define $C = rY - sZ$ for arbitrary constants r and s. The results for this generalized outcome variable are given in table 11.

Results

In order to explore the theoretical power of this technique under ideal conditions, we performed a set of simulations of inheritance of a normally distributed X-linked trait. For allele frequencies of either 0.2 or 0.4, parents were chosen at random from a population assumed to be at Hardy-Weinberg equilibrium. The genotypes of two offspring were generated from these parental genotypes. The mean values of the traits and the residual error variances were chosen to follow either additive, dominant, or recessive models (with respect to females) with heritability proportions of 5, 10, or 20%. In order to get the theoretical best scenario, it was assumed that the IBD sharing could be determined without question. As we chose only one sib pair per pedigree, the problem of within pedigree correlation did not arise.

When a null model was used, the results were very close to the expected type I error rate. Considering three possible outcome variables (squared difference, cross product, and linear combination of squared difference and squared sum weighted with inverse residual variances, respectively), 10,000 replicates of studies with 50 sib pairs each yielded a rejection rate of 4.98, 5.12, and 5.72% respectively. When the number of sib pairs was increased to 500, the rejection rates were 4.63, 5.4, and 5.42% respectively.

The four figures 1–4 show the effects of genetic model and sample size on the theoretical power. In general, a dominant model is the least powerful of the three, and a recessive model is the most powerful of the three, when the frequency of the dominant allele is less than 50%. As expected, the power increases with increasing sample size, but the dominant model is the least susceptible to this effect.

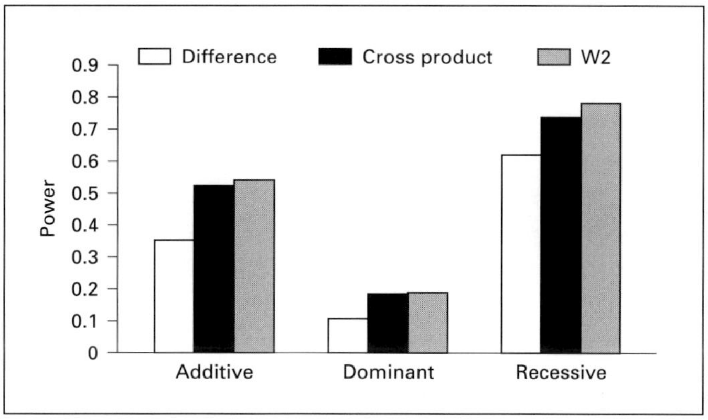

Fig. 1. Effect of genetic model and choice of outcome variable on power. Simulations are based on 500 sib pairs with a locus explaining 20% of the trait variation. The causative allele at this locus has a population frequency of 20%.

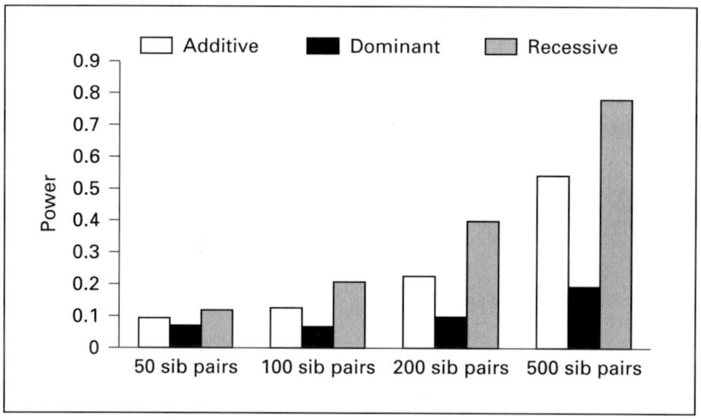

Fig. 3. Effect of sample size and genetic model on power. Simulations are based on a locus explaining 20% of the trait variation. The causative allele at this locus has a population frequency of 20%.

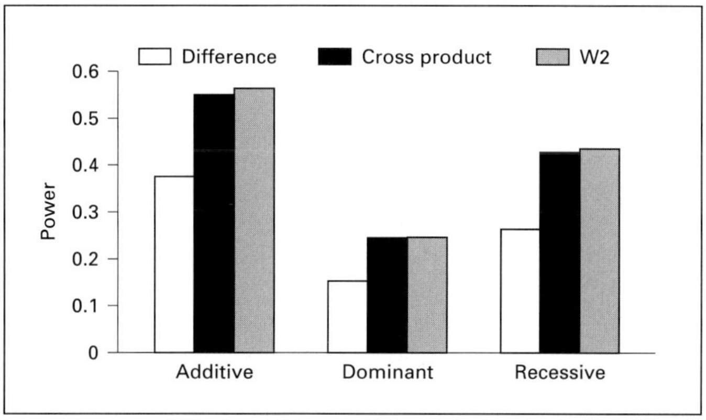

Fig. 2. Effect of genetic model and choice of outcome variable on power. Simulations are based on 500 sib pairs with a locus explaining 20% of the trait variation. The causative allele at this locus has a population frequency of 40%.

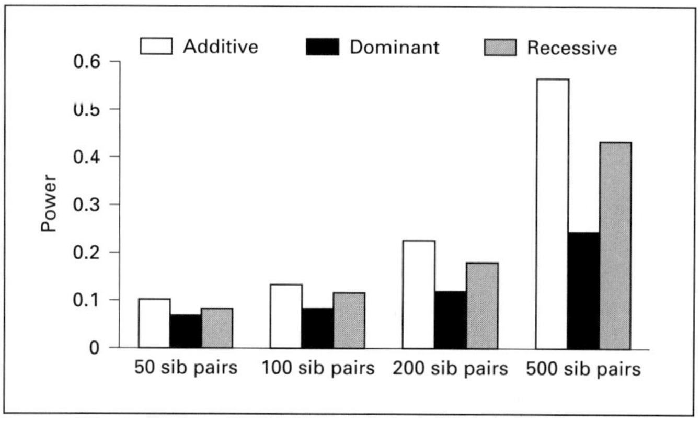

Fig. 4. Effect of sample size and genetic model on power. Simulations are based on a locus explaining 20% of the trait variation. The causative allele at this locus has a population frequency of 40%.

Discussion

Several nonparametric methods have been developed to analyze pedigree data for X linkage. For a good exposition and power comparisons see Song et al. [2002]. Also, Cordell et al. [1995] have given an X chromosome extension of Risch's Maximum Lod Score method for affected relative pair analysis [Risch, 1990a, b, c]. But the X-linked extension of the H-E method has never been documented in the literature. In this manuscript, we have given a theoretical description of the X-linked version of H-E and its revised version combining both the difference-squared and the sum-squared with appropriate weights. It is evident from the results that the method has virtually no power unless the mode of inheritance is recessive. So far we have only considered random samples, which we expect to have less power than an appropriately selected sample. Therefore, we intend extending this work to the case of selected samples similar to the work done for autosomal loci, such as selecting individuals from the tails of a distribution [Carey and Williamson, 1991], or extreme discordant sib pairs [Risch and Zhang, 1995, 1996], which can be expected to increase the power to detect linkage to the X chromosome. Multipoint versions of X linkage and linkage to pseudoautosomal regions [Dupuis and van Eerdewegh, 2000] using sib pairs in the context of the H-E method will be the focus of future research. The X-linked extension discussed above and its multipoint version will be implemented in the computer program SIBPAL [S.A.G.E., 2002].

Acknowledgements

Dr. Wiener is supported by a grant R01 MH66181-01 from NIH/NIMH and R01 NS45934-05 NIH/NINDS awarded to Dr. Go. Dr. Elston is supported in part by US Public Health Service research grant GM 28365 from the National Institute of General Medical Services, and resource grant RR 03655 from the National Center for Research Resources. Dr. Tiwari is supported in part by 1R01AG1892201 (PI: Weindruch) and P20RR016430 (PI: Mohatt) from NIH/NIA and NIH/NCRR, respectively. We thank two anonymous reviewers for a critical review of the manuscript.

Appendix

Estimation of IBD Sharing among Siblings

In order to estimate IBD sharing, we will start by enumerating every possible type of sibship. Recall the four possible mating types, as given in table 12. Because the father is hemizygous, there are only four possible mating types, as opposed to the seven possible when considering an autosomal locus. On the other hand, the number of possible types of sibships will be much larger than those possible at an autosomal locus because we will make the distinctions among all female, all male, and mixed gender sibships. When giving a sibship type below, we understand a stated genotype to indicate any number of siblings with that genotype.

In order to determine the possible sibship types that each mating type can give rise to, we first list the possible genotypes of children that can come from each mating type. For example, from a type III mating, the possible children are daughters with genotypes ii and ij, and sons with genotypes i and j. The possible sibships that arise from these possibilities are given by taking every possible combination. Thus, one possible sibship is all homozygous daughters {ii}; another is all heterozygous daughters {ij}; another is all genetically identical (at that locus) sons {i}; another is a mixture of homozygous and heterozygous daughters {ii,ij}, etc. Table 13 summarizes all of the possible sibship types, and the mating types that can give rise to them. All the possible offspring from a given mating type are equally likely. The probability of each of the four offspring listed from a type III mating is therefore ¼. The probability of a given sibship with N siblings is therefore given by a multinomial distribution. In the case of type I and type II matings, this reduces to a binomial distribution, as only two offspring are possible (one son and one daughter in each case). This multinomial distribution gives the probability of a given sibship conditional on the mating type. This conditional probability is extremely important for the estimation of IBD.

In cases where the parental genotypes are not known – if the parents were unavailable for the study or unwilling to participate – there are not only multiple parental genotypes that can give rise to the observed sibships, but also, as seen in table 13, multiple parental mating types that can give rise to the observed sibship type. Table 14 lists the possible parental genotypes that can give rise to any particular observed sibship type, listed separately by parental mating type. In the following table, an allele listed as '0' means an allele that is arbitrary, subject to the definition of the mating type. For example, a type III parental mating i0 × i can give rise to a sibship of all daughters with genotype ii, but the mating type is only type III if '0' represents any allele other than allele 'i'. Even though the type I mating ii × i is subsumed by the more general i0 × i, the distinction is made here because the probability of the parental genotype, the conditional probability of the sibship given the parental genotype, and the expected IBD value differ by mating type. If two distinct parental genotypes of the same mating type can give rise to the same observed sibship type, a semicolon in table 14 separates the paternal genotypes.

Note that the mating type is unambiguously identified for 12 of the 18 sibship types, and that the parental genotypes are unambiguously defined for all but 3 of these 12. The parental genotype marked with an asterisk (*) is given as i0 × 0′. In this case, the '0' indicates any allele except the 'i' allele. The '0′' indicates any allele

Table 12. Definition of mating types

Mating type	Definition
I	ii × i
II	ii × j
III	ij × i
IV	ij × k

Table 13. Mating types that give each sibship type

Sibship type		Mating type(s)
1	{ii}	I, III
2	{ij}	II, III, IV
3	{ii,ij}	III
4	{ik,jk}	IV
5	{i}	I, II, III, IV
6	{i,j}	III, IV
7	{ii,i}	I, III
8	{ij,i}	II, III, IV
9	{ii,j}	III
10	{ij,k}	IV
11	{ii,ij,i}	III
12	{ii,ij,j}	III
13	{ik,jk,i}	IV
14	{ii,i,j}	III
15	{ij,i,j}	III
16	{ij,i,k}	IV
17	{ii,ij,i,j}	III
18	{ij,ik,j,k}	IV

Table 14. Possible parental genotypes for each sibship type

Sibship type		For mating type:			
		I	II	III	IV
1	{ii}	ii × i		i0 × i	
2	{ij}		ii × j; jj × i	ij × i; ij × j	i0 × j; j0 × i
3	{ii,ij}			ij × i	
4	{ik,jk}				ij × k
5	{i}	ii × i	ii × 0	i0 × i; i0 × 0	i0 × 0′ (*)
6	{i,j}			ij × i; ij × j	ij × 0
7	{ii,i}	ii × i		i0 × i	
8	{ij,i}		ii × j	ij × i; ij × j	i0 × j
9	{ii,j}			ij × i	
10	{ij,k}				ik × j; jk × i
11	{ii,ij,i}			ij × i	
12	{ii,ij,j}			ij × i	
13	{ik,jk,i}				ij × k
14	{ii,i,j}			ij × i	
15	{ij,i,j}			ij × i; ij × j	
16	{ij,i,k}				ik × j
17	{ii,ij,i,j}			ij × i	
18	{ij,ik,j,k}				jk × i

Table 15. IBD probabilities for all sib pair types by mating type

Sib pair type		For mating type:			
		I	II	III	IV
1	{ii,ii}	0.5; 0.5		0; 1	
2	{ij,ij}		0.5; 0.5	0; 1	0; 1
3	{ii,ij}			1; 0	
4	{ik,jk}				1; 0
5	{i,i}	0.5; 0.5	0.5; 0.5	0; 1	0; 1
6	{i,j}			1; 0	1; 0
7	{ii,i}	0.5; 0.5		0; 1	
8	{ij,i}		0.5; 0.5	1; 0 or 0;1	0; 1
9	{ii,j}			1; 0	
10	{ij,k}				1; 0

other than either the 'i' allele, or the allele in place of the '0'. The following table summarizes the probabilities of the levels of IBD for each possible type of sibling pairs for each mating type. Recall that there are only two possible values of IBD for a given sibling pair: 1 and 2 for sisters, 0 and 1 for other gender combinations. The values in table 15 are the probabilities of the two values of IBD. For the case of a type 8 sibling pair, there were two distinct type III matings that could have given rise to that pair. The expected IBD is different for these two matings. In the other cases of differing parental genotypes of the same mating type giving rise to the same sibling pair type (e.g. 2/II, 5/III, etc.), the expected IBD is the same for all possible parental genotypes.

If the parental genotypes are known, we need only look up the values in table 15 to calculate the estimated IBD. For example, if the mating type is known to be type II, and the two siblings in a sibling pair both have genotype ij (i.e. sibling pair type 2), we see in table 4 that $P(\Omega = 1) = P(\Omega = 2) = 0.5$, so the estimate of IBD is $(0.5 \times 1) + (0.5 \times 2) = 1.5$. If the parental data are only partially known, or completely unknown, the estimated IBD at a locus between two siblings will be a weighted average of the expected IBD for each possible parental genotype. The weights are the probabilities of the parental genotypes. These are given as follows: $P(\text{ii} \times \text{i}) = p_i^3$; $P(\text{ii} \times \text{j}) = p_i^2 p_j$; $P(\text{ij} \times \text{i}) = 2p_i^2 p_j$, and finally $P(\text{ij} \times \text{k}) = 2p_i p_j p_k$. When the parental genotype that can give rise to a particular sibship type includes an arbitrary allele, i.e. those listed with '0' alleles in table 14, the weighting factor for that group of genotypes is a sum. For example, the probability of genotype i0 × i with mating type III is given as

$$\sum_{j \neq i} 2p_i^2 p_j = 2p_i^2 \sum_{j \neq i} p_j = 2p_i^2(1 - p_i) = 2(p_i^2 - p_i^3)\,.$$

This is the probability of having a type III parental mating giving rise to a type 1 sibship. Table 16 summarizes the data contained in tables 14 and 15. Each nonempty cell of the table contains the probability of the possible parental genotype, and the estimated IBD for the given sibship type and parental mating type. The estimated IBD value is only given for types 1 through 10, since the remaining sibship types require at least three genotypically different siblings, and pairs of siblings from a sibship will have to be one of the first ten types.

To see how to use these tables to determine IBD among siblings at a locus, consider the following situation. Let's say that the locus has four alleles with frequencies p_1, p_2, p_3, and p_4. One pedigree in our study is a nuclear family with three siblings that consent to the study, but one parent has already died, and the other is unwilling to participate. We thus have the situation of a sibship with no known parental data. The sibship consists of two sisters and a brother. The genotypes

Table 16. Parental genotype probabilities, and estimated IBD

Sibship type		For mating type:			
		I	II	III	IV
1	{ii}	$p_i^3 \Rightarrow 1.5$		$2p_i^2(1-p_i) \Rightarrow 2$	
2	{ij}		$p_i^2 p_j \Rightarrow 1.5$	$2p_i^2 p_j \Rightarrow 2$	$2p_i p_j(1-p_i-p_j) \Rightarrow 2$
			$p_i^2 p_i \Rightarrow 1.5$	$2p_i p_j^2 \Rightarrow 2$	
3	{ii,ij}			$2p_i^2 p_j \Rightarrow 1$	
4	{ik,jk}				$2p_i p_j p_k \Rightarrow 1$
5	{i}	$p_i^3 \Rightarrow 0.5$	$p_i^2(1-p_i) \Rightarrow 0.5$	$2p_i^2(1-p_i) \Rightarrow 1$	$(1-p_i)^2 - \sum_{j \neq i} p_j^2 \Rightarrow 1$
				$2p_i \sum_{j \neq i} p_j^2 \Rightarrow 1$	
6	{i,j}			$2p_i^2 p_j \Rightarrow 0$	$2p_i p_j(1-p_i-p_j) \Rightarrow 0$
				$2p_i p_j^2 \Rightarrow 0$	
7	{ii,i}	$p_i^3 \Rightarrow 0.5$		$2p_i^2(1-p_i) \Rightarrow 0$	
8	{ij,i}		$p_i^2 p_j \Rightarrow 0.5$	$2p_i^2 p_j \Rightarrow 0$	$2p_i p_j(1-p_i-p_j) \Rightarrow 1$
				$2p_i p_j^2 \Rightarrow 1$	
9	{ii,j}			$2p_i^2 p_j \Rightarrow 0$	
10	{ij,k}				$2p_i p_j p_k \Rightarrow 0$
11	{ii,ij,i}			$2p_i^2 p_j$	
12	{ii,ij,j}			$2p_i^2 p_j$	
13	{ik,jk,i}				$2p_i p_j p_k$
14	{ii,i,j}			$2p_i^2 p_j$	
15	{ij,i,j}			$2p_i^2 p_j$	
				$2\, p_i p_j^2$	
16	{ij,i,k}				$2p_i p_j p_k$
17	{ii,ij,i,j}			$2p_i^2 p_j$	
18	{ij,ik,j,k}				$2p_i p_j p_k$

In this table, the number after the sign ⇒ is the MLE of average IBD sharing for the given sibling pair type conditional on the given mating type.

of the sisters at the locus of interest are 1|2 and 1|3, and the genotype of the brother is 3. We identify this sibship as type 13 with i = 3, j = 2, and k = 1. There is only one parental configuration that could have given rise to these children; a mother of type 2|3, and a father of type 1. From table 16, the probability of this parental genotype is $2p_1p_2p_3$. The three possible sib pairs from this trio are the two sisters 1|2 and 1|3 (type 4), the brother-sister pair of types 1|2 and 3 (type 10), and the brother sister pair with types 1|3 and 3 (type 8). From table 16, the estimated IBD for the first pair is $2p_1p_2p_3 \times 1$. Also from table 16, the estimated IBD for the second pair is $2p_1p_2p_3 \times 0 = 0$. When we wish to use table 16 to estimate IBD for the third pair, we see that there are three different possibilities for the genotypes of parents that could have given rise to that type of sibling pair. However, we already know from the full sibship that only one of the three possibilities, the type IV mating, represents the true situation. Therefore, instead of summing over all possible matings that could give rise to that sib pair, we use only the information in the column for the type IV mating. From table 16, we might think that the estimated IBD for the third pair is $2p_1p_3(1 - p_1 - p_3) \times 1$, but it is actually $2p_1p_2p_3 \times 1$ because the knowledge of the full sibship limits the possible parental genotypes. In other words, when we choose a weight, we always use the weight determined by the full sibship (in this case a type 13 sibship) rather than the weights determined only by the sib pair for which we are estimating the IBD.

The values calculated above are the expected values of IBD based on the joint probability of IBD and the parental genotypes. We divide these by the probability of the parental genotype based on the observed data in the pedigree, which is the term shown above: $2p_1p_2p_3$. Thus, the estimates of IBD are simply 1, 0, and 1 respectively. In this particular case, because only one parental genotype configuration was possible, it was not necessary to use the probability of that genotype, as it just cancelled out at the end. In general, more than one such configuration may be possible. When there is more than one such configuration, we calculate the terms as in the previous paragraph – complete with the probabilities of the parental genotypes – and sum them. This sum is then divided by the sum of the probabilities of the parental genotypes; i.e. it is simply a weighted average of the IBD estimate terms from table 16, with weights also given in table 16. This technique will also work if one of the parental genotypes is known. In that case, all of the weights in both the numerator and denominator will include the probability of the observed genotype from the known parent. Thus, the known genotype term(s) will cancel out, reducing to a weighted average over the possible genotypes of the unknown parent.

References

Abecasis GR, Cherny SS, Cookson WO, Cardon LR: 'Merlin-rapid analysis of dense genetic maps using sparse gene flow trees', Nat Genet 2002;30:97–101.

Almasy L, Blangero J: 'Multipoint quantitative-trait linkage analysis in general pedigrees', Am J Hum Genet 1998;62:1198–1211.

Blangero J, Almasy L: 'SOLAR: sequential oligogenic linkage analysis routines', Population Genetics Laboratory Technical Report No. 6. San Antonio, Southwest Foundation for Biomedical Research, 1996.

Carey G, Williamson J: Linkage analysis of quantitative traits: Increased power by using selected samples. Am J Hum Genet 1991;49:786–796.

Cordell HJ, Kawaguchi Y, Todd JA, Farrall M: An extension of the maximum lod score method to X-linked loci. Ann Hum Genet 1995;59:435–449.

Drigalenko E: How sib pairs reveal linkage. Am J Hum Genet 1998;63:1242–1245.

Dupuis J, van Eerdewegh P: Multipoint linkage analysis of the pseudoautosomal regions, using affected sibling pairs. Am J Hum Genet 2000; 67:462–475.

Elston RC, Buxbaum S, Jacobs KB, Olson JM: Haseman and Elston revisited. Genet Epidemiol 2000;19:1–17.

Forrest WF: Weighting improving the 'New Haseman-Elston' method. Hum Hered 2001;52:47–54.

Gudbjartsson DF, Johansson K, Frigge ML, Kong A: Alegro, a new computer program for multipoint linkage analysis. Nat Genet 2000;25:12–13.

Haseman JK, Elston RC: The investigation of linkage between a quantitative trait and a marker locus. Behav Genet 1972;2:3–19.

Kruglyak L, Daly MJ, Reeve-Daly MP, Lander ES: Parametric and nonparametric linkage analysis: a unified multipoint approach. Am J Hum Genet 1996;58:1347–1363.

Risch N: Linkage strategies for genetically complex traits. I. Multilocus models. Am J Hum Genet 1990a; 46:222–228.

Risch N: Linkage strategies for genetically complex traits. II. The power of affected relative pairs. Am J Hum Genet 1990b; 46:222–228.

Risch N: Linkage strategies for genetically complex traits. III. The effect of marker polymorphism on analysis of affected relative pairs. Am J Hum Genet 1990c; 46:229–241.

Risch N, Zhang, H: Extreme discordant sib pairs for mapping quantitative trait loci in humans. Science 1995;268:1584–1589.

Risch N, Zhang H: Mapping quantitative trait loci with extreme discordant sib pairs: Sampling considerations. Am J Hum Genet 1996;58: 836–843.

S.A.G.E.: Statistical Analysis for Genetic Epidemiology, Release 4.3. Computer program available from Statistical Solutions Ltd, Cork, Ireland, 2002.

Sham PC, Purcell S: Equivalence between Haseman-Elston and variance-components linkage analyses for sib pairs. Am J Hum Genet 2001: 68:1527–1532.

Sham PC, Purcell S, Cherny SS, Abecasis GR: Powerful regression-based quantitative-trait linkage analysis of general pedigrees. Am J Hum Genet 2002;71:238–253.

Shete S, Jacobs KJ, Elston RC: Adding further power to the Haseman and Elston method for detecting linkage. Genet Epidemiol 1999;17: 194.

Song KK, Feingold E, Weeks DE: Statistics for Nonparametric Linkage Analysis of X-linked Traits in General Pedigrees, Am J Hum Genet 2002;70:181–191.

Original Paper

Human Heredity

Hum Hered 2003;55:108–116
DOI: 10.1159/000072315

Effect of Box-Cox Transformation on Power of Haseman-Elston and Maximum-Likelihood Variance Components Tests to Detect Quantitative Trait Loci

C.J. Etzel[a] S. Shete[a] T.M. Beasley[b] J.R. Fernandez[b,c,d] D.B. Allison[b,c,d] C.I. Amos[a]

[a]Department of Epidemiology, University of Texas, M.D. Anderson Cancer Center, Houston, Tex.;
[b]Department of Biostatistics, Section on Statistical Genetics, [c]Department of Nutrition Sciences, Division of Physiology and Metabolism, and Clinical Nutrition Research Center, University of Alabama at Birmingham, Birmingham, Ala., USA

Key Words

Box-Cox transformation · Kurtosis · Sib pair linkage analysis · Skewness · Non-normality · Winsorizing

Abstract

Non-normality of the phenotypic distribution can affect power to detect quantitative trait loci in sib pair studies. Previously, we observed that Winsorizing the sib pair phenotypes increased the power of quantitative trait locus (QTL) detection for both Haseman-Elston (HE) least-squares tests [Hum Hered 2002;53:59–67] and maximum likelihood-based variance components (MLVC) analysis [Behav Genet (in press)]. Winsorizing the phenotypes led to a slight increase in type 1 error in H-E tests and a slight decrease in type I error for MLVC analysis. Herein, we considered transforming the sib pair phenotypes using the Box-Cox family of transformations. Data were simulated for normal and non-normal (skewed and kurtic) distributions. Phenotypic values were replaced by Box-Cox transformed values. Twenty thousand replications were performed for three H-E tests of linkage and the likelihood ratio test (LRT), the Wald test and other robust versions based on the MLVC method. We calculated the relative nominal inflation rate as the ratio of observed empirical type 1 error divided by the set α level (5, 1 and 0.1% α levels). MLVC tests applied to non-normal data had inflated type I errors (rate ratio greater than 1.0), which were controlled best by Box-Cox transformation and to a lesser degree by Winsorizing. For example, for non-transformed, skewed phenotypes (derived from a χ^2 distribution with 2 degrees of freedom), the rates of empirical type 1 error with respect to set α level = 0.01 were 0.80, 4.35 and 7.33 for the original H-E test, LRT and Wald test, respectively. For the same α level = 0.01, these rates were 1.12, 3.095 and 4.088 after Winsorizing and 0.723, 1.195 and 1.905 after Box-Cox transformation. Winsorizing reduced inflated error rates for the leptokurtic distribution (derived from a Laplace distribution with mean 0 and variance 8). Further, power (adjusted for empirical type 1 error) at the 0.01 α level ranged from 4.7 to 17.3% across all tests using the non-transformed, skewed phenotypes, from 7.5 to 20.1% after Winsorizing and from 12.6 to 33.2% after Box-Cox transformation. Likewise, power (adjusted for empirical type 1 error) using leptokurtic phenotypes at the 0.01 α level ranged from 4.4 to 12.5% across all tests with no transformation, from 7 to 19.2% after Winsorizing and from 4.5 to 13.8% after Box-Cox transformation. Thus the Box-Cox transformation apparently provided the best type 1 error con-

KARGER

Fax + 41 61 306 12 34
E-Mail karger@karger.ch
www.karger.com

0001–5652/03/0553–0108$19.50/0

Accessible online at:
www.karger.com/hhe

C.J. Etzel
Department of Epidemiology, University of Texas
M.D. Anderson Cancer Center
1515 Holcombe Blvd, Box 189, Houston, TX 77030 (USA)
Email cetzel@manderson.org

trol and maximal power among the procedures we considered for analyzing a non-normal, skewed distribution (χ^2) while Winzorizing worked best for the non-normal, kurtic distribution (Laplace). We repeated the same simulations using a larger sample size (200 sib pairs) and found similar results.

Introduction

The ability to detect linkage to quantitative trait loci (QTLs) using model-based and model-free methods may be mediated by the distribution of the phenotype under study. Linkage methods based on sib pairs assess a measure of allele sharing, typically identity by descent (IBD) with respect to phenotypic similarity. The distribution of the measure of phenotypic similarity may further affect linkage detection. Included in the class of model-free sib pair linkage methods are the Haseman-Elston method [3] and MLVC method [4, 5]. The original Haseman-Elston (OHE) method is an ordinary least-squares regression-based QTL linkage method in which a phenotypic similarity measure, the squared sib pair difference $Y_j = (X_{1j} - X_{2j})^2$, is regressed onto the proportion of alleles π_j shared IBD. This OHE method was revised [6] and a new phenotypic similarity measure was proposed, the sib pair mean-corrected cross-product, $Z_j = (X_{1j} - \bar{X})(X_{2j} - \bar{X})$ where $\bar{X}$ is the sample mean for all sibs. The sib pair mean-corrected cross-product is then regressed onto π_j. We refer to this method as the revised Haseman-Elston (RHE). An additional Haseman-Elston procedure (NHE) has been proposed [7] in which a weighted linear combination of the squared sib pair difference (Y_D) and square mean-corrected sib pair sum (Y_S) is regressed onto π, such that

$$W_j = \frac{c_S Y_{Sj} + c_D Y_{Dj}}{c_S + c_D}$$

where

$$Y_{Sj} = 0.25[(X_{1j} - \bar{X})(X_{2j} - \bar{X})]^2$$

and

$$Y_{Dj} = 0.5(X_{1j} - X_{1j})^2 .$$

The weights c_S and c_D are given by

$$c_S = (1 - r_S)\sigma_S^2 \frac{N}{N-2}$$

and

$$c_D = (1 - r_D)\sigma_D^2 \frac{N}{N-2},$$

where $r_.$ is the correlation between π and $Y_.$, $\sigma_.^2$ is the variance of $Y_.$ and N is the number of sib pairs. For specific values of c_S and c_D, NHE reduces to OHE and RHE [1]. Variations of this NHE have been independently published by several groups [8–12].

MLVC analysis models the expected covariance between two related individuals' phenotypes given π as a function of major gene (σ_a^2), polygene (σ_G^2) and environment (σ_e^2) variance components. A likelihood ratio test (LRT) can be used to assess linkage. The LRT is calculated as $-2\ln(L_0/L_1)$ where L_1 is the likelihood under the unrestricted model where σ_a^2, σ_G^2 and σ_e^2 are estimated and L_0 is the likelihood under the restricted model where $\sigma_a^2 = 0$. Linkage can also be tested using a Wald test in which the maximum-likelihood estimate of additive variance is divided by its standard error. Amos et al. [13] developed a robust Wald test (RWald) based upon a sandwich estimator of the variance [14].

The various HE procedures are attractive methods because they are simple and can be completed using standard regression software packages. However, power to detect linkage using the OHE is relatively low except in specific cases [15]. Under the assumption of equal sib pair means, $E(Z_j)$ from the RHE is the sibling covariance. In circumstances of moderate to high among-sib phenotypic correlation, the RHE regression slope estimate is not efficient [11]. When non-normal sib pair phenotypes are analyzed, the functions of phenotypic similarity from the OHE and the NHE further exacerbate skewness. The RHE and NHE have been shown to increase power compared with the OHE to detect linkage while still maintaining an appropriate type 1 error rate [7, 16]. However, power to detect linkage under non-normal phenotypic distributions is still adversely affected [1]. MLVC methods are widely used because multiple genetic and environmental effects can be modeled. However, some distribution (usually normality) must be assumed for maximum-likelihood estimation and testing in MLVC analysis. Allison et al. [17] showed that non-normally distributed phenotypes lead to excessive type 1 error rates. Blangero et al. [18] proposed a kurtosis correction to the likelihood ratio (KLRT) to eliminate the excessive type 1 error rate of the MLVC. The KLRT procedure adjusts the χ^2 value for detecting linkage, subtracting a quantity that depends upon the major-gene-related heritability for positive kurtosis [2]. Because the major gene heritability is ordinarily unknown, a more conservative procedure than trying to fit components of kurtosis due to environmental and genetic causes as suggested by Blangero et al. [18] adjusts for the overall heritability. However, even if the kurtosis adjustment is arbitrarily confined to the genetic components, there is a choice of whether to include the broad heritability in the adjustment, which we denote KLRT1 or the locus-specific heritability which we denote KLRT2. We observed inflated type 1 error rates for Winsorized data using the kurtic correction to the LRT [2]. The RWald test is robust to model misspecification [19]. However the Wald and RWald test have inflated type 1 error for non-normal data [2] for sample sizes of 100 sib pairs.

When using statistical procedures that are based on normality assumptions, it is common to transform the non-normal data prior to analysis. The Box-Cox transformation and other monotone transformations toward normality have been studied in genetic epidemiology with mixed effectiveness. Because segregation of a major gene imparts platykurtosis and may impart skewness, transformation of the data prior to analysis may reduce the power to detect a genetic factor [20] while protecting against false positive inferences, but some of the tests that are used to identify genetic factors often have higher power if the data are approximately normally distributed [21, 22]. Commingling analysis and similar methods like segregation analysis, which rely heavily upon distributional assumptions concerning the trait distribution, can be strongly influenced by the type of transformation (if any) that is used [23, 24]. On the other hand, model-free linkage analysis, which correlates marker IBD with trait values is not expected to be as heavily influenced by the transformation that is used. However, maximum-likelihood based tests require that the data conform with the analytical model that is used. For example, when a normal model is postulated for the distribution of trait values and maximum-likelihood based variance components tests are applied to non-normal data, Allison et al. [17] showed a very marked increase in type 1 error rates, when the siblings showed a residual correlation. Then again, regression-based linkage tests can use the central limit theorem to ensure that the test statistics that are constructed follow a normal distribution, and so these methods are typically well behaved when the data are non-normal.

Blackwelder and Elston [15] found no benefit in the OHE after power and log transformations of squared sib pair differences prior to regression. However, in a previous study [1], we observed an increase in power for the OHE when the individual sibling phenotypes were transformed using Winsorizing prior to the squared sib pair calculation. We also observed an increase in power for the RHE, NHE and LRT, Wald and other robust MLVC measures [2] when the individual sibling phenotypes were Winsorized. Winsorizing truncates the distribution at pre-determined percentiles and hence would be an appropriate transformation for distributions that are slightly skewed and/or kurtic. Winsorizing may not be the best transformation to normality for highly skewed distributions. Instead, the Box-Cox family of transformations may be more appropriate for highly skewed distributions. Herein, we compare the effects of Winsorizing and Box-Cox transformation on power and type 1 error for the three HE tests and the MLVC tests.

Methods

We completed two simulation series. In series 1, we generated 20,000 replicates with 100 sib pairs per replicate to investigate the effect of transformation on type 1 error. In series 2, we generated 10,000 replicates with 200 sib pairs per replicate to compare power. In both simulation series, the phenotype value for each sib consisted of mean displacement due to the QTL and a residual value. Sibling residual values were correlated. The residual value for each sibling in a pair was simulated from a standard normal, χ^2 with 2 degrees of freedom and a central Laplace (i.e., double exponential distribution [25]) with variance of 8. The χ^2 distribution has a skewness of 2 and kurtosis of 6. The Laplace has a skewness of 0 and kurtosis of 3. The residual sib pair correlation was 0.50. Excess in the rate of type 1 error for MLVC procedures [15] but not Haseman-Elston tests [16] has been reported for these two non-normal distributions. The description of the simulation of the sib pair data is detailed in Fernandez et al. [1]. Briefly, we first generated the residual value for sib 1 and then generated the residual value for sib 2 such that the marginal distributions were identical and correlation between the two residuals was 0.50. Then the mean displacement due to the QTL was added to each sibling's phenotype value. The resulting phenotypic values for sib 1 and sib 2 in the ith sib pair are thus denoted as X_{i1} and X_{i2}, respectively. The phenotype values were then scaled to have zero mean and unit variance. The Box-Cox transformation requires that $X_{ij}>0$, so we shifted the standardized distribution of X by adding the absolute value of the min(X) plus a small positive constant to each value.

The individual phenotypes were then transformed by Winsorizing or Box-Cox transformation. The Winsorized set $X_W = \{X_{Wij}: i = 1,..., k; j = 1, 2\}$ was constructed as

$$X_{Wij} = \begin{cases} C_\zeta, \text{ for } X_{ij} \leq C_\zeta \\ X_{ij} \text{ for } C_{1-\zeta} > X_{ij} > C_\zeta \\ C_{1-\zeta}, \text{ for } X_{ij} > C_{1-\zeta} \end{cases}$$

where k is the number of sib pairs and C_ζ and $C_{1-\zeta}$ are the ζ100th and $(1 - \zeta)$100th percentiles of the distribution of X. We chose ζ = 0.05 and reassigned values that were not within the 5th and 95th percentiles. The Box-Cox transformed set $X_B = \{X_{Bij}: i = 1,..., k; j = 1, 2\}$ was constructed as

$$X_{Bij} = \begin{cases} \dfrac{X_{ij}^\lambda - 1}{\lambda}, \text{ for } \lambda \neq 0 \\ \ln X_{ij}, \text{ for } \lambda = 0 \end{cases}$$

For each simulation replicate, the choice of λ was the value that maximized the log-likelihood from a normal distribution given by

$$L(\lambda) = -\frac{2k}{2}\ln\left[\frac{1}{2k}\sum_{i=1}^{k}\sum_{j=1}^{2}(X_{Bij}-\overline{X}_B)^2\right] + (\lambda - 1)\sum_{i=1}^{k}\sum_{j=1}^{2}\ln X_{Bij}$$

where

$$\overline{X}_B = \frac{1}{2k}\sum_{i=1}^{k}\sum_{j=1}^{2}X_{Bij}.$$

The OHE, RHE and NHE tests were performed on the non-transformed, Winsorized and Box-Cox transformed data from each of the three distributions. We also completed five MLVC tests: LRT, Wald, RWald, KLRT1 that uses broad sense heritability, KLRT2 that uses locus-specific sense of heritability (KLRT2). Type 1 error was evaluated for α values of 0.05, 0.01 and 0.001. We constructed 95% confidence intervals as

$$\hat{\alpha} \pm 2\sqrt{\frac{\alpha(1-\alpha)}{R}},$$

where $\hat{\alpha}$ is the observed type 1 error, α is the true type 1 error rate and R is the number of replications. An observed type 1 error rate is defined as *conservative* if α is less than the lower bound of the confidence interval and *inflated* if α is greater than the upper bound of the confidence interval. Because the different tests that we considered have different type 1 errors, we used the empirical distribution of the observed p values from the type 1 error simulations to define a cut-off point to calculate power for each of the tests. This enabled us to directly compare the power for the eight tests of linkage. We further conducted tests of significance using a two-sided test for proportions to determine if there was a significant change in power after transformation and between transformations within linkage tests for each phenotype.

Results

Table 1 displays the type 1 error rates for the simulated linkage tests from series 1 using non-transformed, Winsorized and Box-Cox transformed data as the ratio of observed empirical p value to α level for α values of 0.05, 0.01 and 0.001. The confidence intervals for these ratios were not provided due to space constraints. However, these can be obtained by converting the ratios back to empiric type 1 error rates and noting that the type 1 error estimate follows a binomial distribution. Where applicable, these ratios are labeled as conservative ($^+$) or inflated (*). The OHE attained expected type 1 error rate in most simulations except at the α = 0.001 level where all observed type 1 errors were inflated. The RHE had mostly inflated type 1 error rates at the 0.05 and 0.001 level regardless of the distribution. The NHE either had conservative type 1 error rates or attained the expected α level except for three instances at the 0.001 level where inflated type 1 error was observed. The overall effect of transformation of the three sets of phenotypes on type 1 error was mixed. For the OHE tests neither transformation consistently reduced or inflated the observed type 1 error rates among the three phenotypes. The Box-Cox transformation of the normal and kurtic phenotypes tended to reduce type 1 error rates for the RHE test, while Winsorization of these two distributions lead to mostly inflated error rates. Neither transformation reduced the inflated type 1 errors of the RHE tests performed on the skewed phenotypes. More consistent results were observed for the NHE test where transformation tended to keep the already conservative rates either conservative or within the expected range.

The LRT tended to have inflated type 1 error for the non-normal untransformed phenotypes. Transformation by Winsorizing tended to increase the type 1 error of the normal phenotypes at the 0.05 and 0.01 level and decreased the type 1 error rate at the 0.001 level. In contrast, Box-Cox transformation markedly decreased the substantially inflated empirical type 1 error of the LRT for the χ^2 and Laplace data sets. The KLRT1 and KLRT2 tests had inflated empirical type 1 error for the normal, non-transformed data and conservative type 1 error rates for non-normal, non-transformed data at the 0.05 level. Winsorizing substantially inflated empirical error rates for the KLRT tests. After Box-Cox transformation, the empirical type 1 error remained inflated for the normal and skewed phenotypes, but became conservative for the kurtic phenotypes. At the 0.01 and 0.001 levels, the KLRT tests had expected or slightly conservative type 1 error for all non-transformed phenotypes. Winsorizing led to inflated type 1 error rates for all phenotypes for the KLRT tests. For the KLRT tests, the type 1 error rates at the 0.01 and 0.001 levels for the Box-Cox transformed phenotypes had smaller type 1 error rates compared to Winsorized phenotypes, although the χ^2 phenotypes still had inflated error rates. The empirical type 1 error was most inflated for the two Wald tests. Both transformations tended to reduce these rates. However, the Wald tests still gave inflated nominal rates. In our simulations involving 200 sib pairs (data not shown), the type 1 error rate of the RWald test showed a closer conformance with the nominal type 1 error rate than it had for 100 sib pairs, and generally showed slightly lower type 1 error rates than the Wald test. At a nominal 5% significance level, the empiric type 1 error rate was 0.077 for the normal data, 0.090 for the χ^2 distribution and 0.097 for the Laplace distribution, and the behavior of this test was similar for the transformed data. However, even at this sample size the test did not adequately protect against false positive inferences.

Table 1. Type 1 error rate as a relative rate of empirical p value/α for non-transformed, Winsorized and Box-Cox-transformed phenotype data for H-E- and VC-based methods (expected relative ratio = 1.0) from 20,000 replicates with 100 sib pairs per replicate

	Non-transformed			Winsorized			Box-Cox		
	normal	χ^2	Laplace	normal	χ^2	Laplace	normal	χ^2	Laplace
$\alpha = 0.05$									
OHE	0.996	1.033	0.857*	0.917	0.961	0.908*	1.001	1.082$^+$	0.894*
RHE	1.151$^+$	1.360$^+$	1.096$^+$	1.209$^+$	1.290$^+$	1.154$^+$	0.944	1.160$^+$	0.846*
NHE	0.714*	0.813*	0.759*	0.806*	0.850*	0.748*	0.748*	0.825*	0.822*
LRT	1.137$^+$	1.978$^+$	1.833$^+$	1.314$^+$	1.510$^+$	1.682$^+$	1.091$^+$	1.152$^+$	1.808$^+$
KLRT1	1.188$^+$	0.444*	0.769*	2.242$^+$	1.426$^+$	2.007$^+$	1.153$^+$	1.590$^+$	0.642*
KLRT2	1.185$^+$	0.551*	0.848*	2.153$^+$	1.440$^+$	1.949$^+$	1.185$^+$	1.582$^+$	0.919*
Wald	1.628$^+$	2.698$^+$	2.410$^+$	1.718$^+$	2.026$^+$	2.223$^+$	1.572$^+$	1.347	2.319$^+$
RWald	1.954$^+$	2.181$^+$	2.221$^+$	1.893$^+$	1.672$^+$	2.189$^+$	1.859$^+$	1.429$^+$	2.301$^+$
$\alpha = 0.01$									
OHE	1.175$^+$	0.803	1.012	0.955	1.112	1.087	0.175*	0.723	1.146$^+$
RHE	1.081	1.800$^+$	0.986	0.859	1.622$^+$	0.921	0.376*	1.007	0.823
NHE	0.497*	0.832	0.766	1.065	0.590*	0.754	0.653*	0.764	0.836
LRT	0.955	4.350$^+$	3.690$^+$	1.265$^+$	3.095$^+$	2.885$^+$	0.955	1.195$^+$	3.225$^+$
KLRT1	1.110	0.325*	0.855	3.585$^+$	2.635$^+$	3.960$^+$	1.135	2.780$^+$	0.920
KLRT2	1.065	0.585*	0.955	3.180$^+$	2.700$^+$	3.455$^+$	1.085	2.415$^+$	1.010
Wald	1.880$^+$	7.330$^+$	5.885$^+$	2.405$^+$	4.880$^+$	4.915$^+$	1.945$^+$	1.905$^+$	5.725$^+$
RWald	3.750$^+$	4.380$^+$	4.915$^+$	3.375$^+$	3.355$^+$	5.035$^+$	3.360$^+$	2.765$^+$	4.570$^+$
$\alpha = 0.01$									
OHE	2.000$^+$	2.030$^+$	1.490$^+$	1.980$^+$	1.820$^+$	1.990$^+$	1.810$^+$	1.940$^+$	1.650$^+$
RHE	1.590$^+$	3.130$^+$	1.770$^+$	1.800$^+$	2.270$^+$	2.410$^+$	1.480$^+$	2.670$^+$	0.590
NHE	0.600	0.680	0.880	2.200$^+$	0.990	2.040$^+$	0.520*	0.910	1.860$^+$
LRT	0.800	14.050$^+$	9.750$^+$	0.650	8.550$^+$	4.000$^+$	0.700	0.650	7.200$^+$
KLRT1	0.850	0.100*	0.150*	4.900$^+$	4.100$^+$	7.550$^+$	0.900	2.250$^+$	0.250*
KLRT2	0.850	0.100*	0.300*	3.450$^+$	6.150$^+$	7.100$^+$	0.850	1.950$^+$	0.400*
Wald	2.950$^+$	37.900$^+$	30.450$^+$	3.850$^+$	20.600$^+$	20.550$^+$	2.950$^+$	3.050$^+$	27.200$^+$
RWald	7.450$^+$	8.800$^+$	14.600$^+$	9.350$^+$	16.850$^+$	14.550$^+$	7.450$^+$	6.200$^+$	14.800$^+$

* Conservative type 1 error; $^+$ inflated type 1 error.

In order to appropriately compare the power between these linkage tests, we used the empirical distribution of the observed p values from the type 1 error simulations to define a cut-off point to calculate power for each of the tests. Table 2 presents the simulation results for the empirical power of the eight linkage tests as adjusted by empirical error rates for series 1 with 100 sib pairs. The values are marked to indicate no significant change in power (at the 5% α level) after transformation (a) or no difference in power between Winsorizing and Box-Cox transformation (b). The RHE had substantially less power than the other tests under most circumstances. The other Haseman-Elston tests had slightly less power than the VCA-based tests. Transformation using Winsorizing and Box-Cox transformation increased the power of the NHE for all phenotypes. Transformation increased power among the non-normal phenotypes for the OHE and NHE tests. The two LRTs that corrected for kurtosis (KLRT1 and KLRT2) had highest power to detect linkage before and after transformation. Generally, the LRT had the next highest power followed by the Wald and RWald tests. For all eight linkage tests, Winsorizing increased power more for the Laplace data while Box-Cox transformation increased power more for the χ^2 data, except for the KLRT tests at the 0.01 or 0.001 levels where Winsorizing increased power more for the χ^2 phenotypes than the Box-Cox transformation.

Table 3 presents the simulation results for the empirical power of the eight linkage tests as adjusted by empirical error rates for series 2 with 200 sib pairs. The values

Table 2. Empirical power for non-transformed, Winsorized and Box-Cox-transformed phenotype data for H-E- and VC-based methods* from 20,000 replicates with 100 sib pairs per replicate

	Non-transformed			Winsorized			Box-Cox		
	normal	χ^2	Laplace	normal	χ^2	Laplace	normal	χ^2	Laplace
$\alpha = 0.05$									
OHE	0.353	0.300	0.229	0.343	0.427	0.340	0.351[a,b]	0.472	0.243
RHE	0.211	0.223	0.179	0.242	0.284	0.229	0.217[a]	0.323	0.183[a]
NHE	0.349	0.344	0.300	0.350[a]	0.432	0.400	0.353[a,b]	0.502	0.317
LRT	0.386	0.273	0.268	0.389[a]	0.371	0.378	0.391[a,b]	0.549	0.292
KLRT1	0.392	0.382	0.376	0.378	0.456	0.427	0.405	0.412	0.395
KLRT2	0.398	0.372	0.359	0.374	0.446	0.413	0.409	0.421	0.383
Wald	0.371	0.234	0.257	0.380[a]	0.337	0.338	0.373[a,b]	0.584	0.252[a]
RWald	0.343	0.302	0.294	0.323	0.362	0.313	0.339[a]	0.558	0.293[a]
$\alpha = 0.01$									
OHE	0.164	0.093	0.057	0.142	0.133	0.159	0.159[a]	0.145	0.070
RHE	0.075	0.099	0.044	0.098	0.131	0.070	0.064	0.126[b]	0.045[a]
NHE	0.153	0.129	0.108	0.154[a]	0.169	0.158	0.165	0.267	0.119
LRT	0.187	0.072	0.076	0.172	0.118	0.119	0.184[a]	0.332	0.081[a]
KLRT1	0.188	0.173	0.125	0.202	0.201	0.192	0.176	0.171[a]	0.138
KLRT2	0.177	0.150	0.123	0.186	0.171	0.179	0.176[a]	0.201	0.129[a]
Wald	0.138	0.047	0.047	0.163	0.075	0.095	0.124	0.308	0.052
RWald	0.128	0.113	0.077	0.113	0.093	0.124	0.136	0.285	0.085
$\alpha = 0.001$									
OHE	0.047	0.028	0.013	0.062	0.007	0.061	0.041	0.083	0.012[a]
RHE	0.028	0.027	0.006	0.030[a]	0.036	0.014	0.028[a,b]	0.052	0.009
NHE	0.056	0.035	0.020	0.064	0.076	0.071	0.056[a]	0.053	0.037
LRT	0.059	0.008	0.010	0.060[a]	0.060	0.042	0.057[a,b]	0.169	0.011[a]
KLRT1	0.073	0.066	0.058	0.059	0.096	0.054[a]	0.066[b]	0.056	0.041
KLRT2	0.069	0.055	0.059	0.065	0.087	0.052	0.063[b]	0.067	0.045
Wald	0.036	0.020	0.012	0.042	0.026	0.022	0.039[a,b]	0.088	0.014[a]
RWald	0.037	0.020	0.010	0.038[a]	0.032	0.030	0.036[a,b]	0.090	0.127

* Adjusted for empirical type 1 error rate observed at the given α level.

[a] Power after transformation not significantly different from power before transformation.

[b] Power after Winsorizing not significantly different from power after Box-Cox transformation.

are marked to indicate no significant change in power (at the 5% α level) after transformation (a) or no difference in power between Winsorizing and Box-Cox transformation (b). Once again, the RHE had substantially less power than the other Haseman-Elston and VCA tests. The OHE and NHE tests had less power than the VCA-based tests and the Wald and RWald tests had the lowest power among all VCA tests. Transformation using Winsorizing and Box-Cox transformation led to significant changes in power among all phenotypes, except for the normal data. Once again, Winsorizing increased power more for the Laplace data while Box-Cox transformation increased power more for the χ^2 data, even for the KLRT tests.

Discussion

We compared type 1 error and power among the H-E tests and VCA methods after transformation of the individual sibling phenotypic values. Winsorizing only minimally decreased the power to detect linkage for the normal distribution using any of the tests. On the other hand, for the Laplace distribution that is highly kurtic, Winsorizing considerably improved the power to detect linkages using any of the test procedures we studied. In contrast, for phenotypic values simulated from a skewed distribution (i.e., χ^2), power to detect linkage was greatest after Box-Cox transformation. The fact that Winsorizing

Table 3. Empirical power for non-transformed, Winsorized and Box-Cox-transformed phenotype data for H-E- and VC-based methods* from 10,000 replicates with 200 sib pairs per replicate

	Non-transformed			Winsorized			Box-Cox		
	normal	χ^2	Laplace	normal	χ^2	Laplace	normal	χ^2	Laplace
$\alpha = 0.05$									
OHE	0.566	0.392	0.394	0.566[a]	0.597	0.614	0.559[a, b]	0.725	0.405
RHE	0.295	0.268	0.322	0.361	0.399	0.464	0.290[a]	0.459	0.334
NHE	0.584	0.483	0.492	0.557	0.635	0.648	0.577[a]	0.786	0.507
LRT	0.603	0.426	0.451	0.557	0.659	0.658	0.576[b]	0.666[b]	0.419
KLRT1	0.606	0.532	0.540	0.575	0.663	0.659	0.572[b]	0.694	0.417
KLRT2	0.604	0.525	0.529	0.572	0.664	0.657	0.571[b]	0.693	0.420
Wald	0.595	0.400	0.426	0.569	0.658	0.652	0.578[b]	0.691	0.412
RWald	0.569	0.438	0.431	0.568	0.632	0.611	0.553	0.697	0.383
$\alpha = 0.01$									
OHE	0.304	0.183	0.191	0.312[a]	0.324	0.352	0.298[a]	0.477	0.202
RHE	0.120	0.091	0.139	0.154	0.176	0.224	0.122[a]	0.215	0.152
NHE	0.321	0.235	0.265	0.286	0.347	0.384	0.317[a]	0.541	0.274
LRT	0.359	0.180	0.188	0.293	0.352	0.387	0.322	0.404	0.226
KLRT1	0.377	0.298	0.284	0.293	0.363	0.380	0.320	0.445	0.217
KLRT2	0.373	0.287	0.280	0.288	0.360	0.382	0.323	0.434	0.219
Wald	0.351	0.144	0.160	0.278	0.329	0.354	0.301	0.445	0.195
RWald	0.309	0.197	0.184	0.257	0.309	0.309	0.273	0.420	0.170
$\alpha = 0.001$									
OHE	0.090	0.049	0.049	0.111	0.088	0.112	0.085[a]	0.255	0.048[a]
RHE	0.034	0.031	0.031	0.038[a]	0.050	0.055	0.033[a]	0.076	0.032[a]
NHE	0.112	0.062	0.057	0.097	0.069	0.158	0.099[b]	0.253	0.057[a]
LRT	0.131	0.036	0.050	0.103	0.065	0.146	0.112	0.168	0.045
KLRT1	0.156	0.129	0.093	0.099	0.095	0.139	0.103[b]	0.193	0.052
KLRT2	0.151	0.107	0.093	0.101	0.094	0.143	0.107[b]	0.188	0.052
Wald	0.146	0.024	0.029	0.074	0.062	0.127	0.068	0.204	0.048
RWald	0.105	0.052	0.041	0.073	0.084	0.114	0.069[b]	0.188	0.044[b]

* Adjusted for empirical type 1 error rate observed at the given α level.
[a] Power after transformation not significantly different from power before transformation.
[b] Power after Winsorizing not significantly different from power after Box-Cox transformation.

worked best for the Laplace distributed data is not surprising because the equally heavy upper and lower tails were reduced by the same percent. For the χ^2 distribution, some skewness may still remain after Winsorizing due to the unequally skewed tails. It is possible to change the value of ζ used in the Winsorizing and hence increase the number of values reassigned in the tail ends of the distribution. However, it has been shown [1] that increasing ζ does not necessarily result in substantial gain of power. For the KLRT tests, transformation by Winsorizing resulted in higher power for the kurtic phenotypes compared to Box-Cox transformation. This result reflects the adjustment by the KLRT tests for kurtosis. Using the two KLRT tests in conjunction with Winsorizing eliminates the influence of possible outliers among the kurtic phenotypes and simultaneously adjusts the distribution of the likelihood ratio tests towards the referent χ^2 distribution.

Transforming non-normal phenotypes prior to analysis is common in genetic studies using regression-based methods, which are most efficient when the errors are normally distributed. Transformation has been shown to increase the power of segregation analysis even in the presence of admixture [26–29]. Power transformations have been used to increase power to detect genetic linkage [30–33]. We applied the Box-Cox family of power transformations and found that power to detect linkage increased for skewed and kurtic phenotypes. However, we observed a greater increase in power from Winsorizing for kurtic phenotypes compared to the power from the Box-Cox transformation.

Here we have evaluated a wide range of tests that can be applied for quantitative data, but additional tests could be considered. Amos et al. [14] suggested a quasi-likelihood procedure that performed better for non-normal data than maximum likelihood methods, but the method requires a Wald-test type of testing scheme. The results from the current studies suggest that type I error rates for Wald tests may not be accurate at the sample sizes we considered. More recently, Sham et al. [9] and Wang [34] have proposed new tests for linkage for quantitative data. The Sham et al. [9] procedure resembles the NHE procedure we studied, but bears further evaluation to compare the power of his approach under different scenarios for the phenotype distributions.

The LRT and KLRT tests provided higher power than Haseman-Elston tests. For the kurtic data, the easily applied kurtosis adjusted likelihood ratio tests provided greater power than other methods. The KLRT tests also provided better power for Winsorized tests, which can be explained by the induced kurtosis from Winsorizing. Generally the KLRT2 test provided a closer conformance to the nominal type I error rate and slightly more power than the KLRT1 test for most situations. The one caveat about these tests based upon modifications of MLVC is the variable type I error rates, which suggests that p values have to be derived from simulation, when reporting results that use these methods. Further studies are needed to evaluate whether asymptotic χ^2 distributions can be applied for the KLRT tests when large sample sizes are studied. For the χ^2 distribution, Box-Cox transformation followed by usual LRT analysis yielded the greatest power. This result is sensible because Box-Cox transformation can be expected to transform the χ^2 distribution to approximate normality. An advantage of the various Haseman-Elston tests is the lack of type I error inflation. Even though they were generally less powerful than the MLVC tests, they are computationally faster and never appear to require simulation to obtain a valid test having the specified type I error rate.

In this study, we focused on power to detect linkage and not estimation. In two previous studies [1, 2], we reported that transformation from Winsorizing leads to increased bias in the estimation of the variance of the major gene for the H-E methods and reduced bias for the MLVC tests. By the nature of the Box-Cox transformation (not location and scale invariant), estimation of the variance of the major gene should be biased and the direction of the bias will be based on the power of the transformation. Therefore, as previously cautioned [1], we recommend that researchers who are focused on estimating QTL effects should exercise care in the interpretation of results from analysis of transformed data.

Acknowledgements

This research was partially supported by funding from NIH grants HG02275 and ES09912.

References

1 Fernandez JR, Etzel CJ, Beasley TM, Shete S, Amos CI, Allison DB: Improving the power of sib pair quantitative trait loci detection by phenotype Winsorization. Hum Hered 2002;53:59–67.
2 Shete S, Beasley TM, Etzel CJ, Fernandez JR, Chen J, Allison DB, Amos CI: Effect of Winsorization on power and type 1 error of variance components and related methods of QTL detection. Behav Genet, in press.
3 Haseman JK, Elston RC: The investigation of linkage between a quantitative trait and a marker locus. Behav Genet 1972;2:3–19.
4 Amos CI: Robust variance-components approach for assessing genetic linkage in pedigrees. Am J Hum Genet 1994;54:535–543.
5 Almasy L, Blangero J: Multipoint quantitative-trait linkage analysis in general pedigrees. Am J Hum Genet 1998;62:1198–1211.
6 Elston RC, Buxbaum S, Jacobs KB, Olson JM: Haseman and Elston revisited. Genet Epidemiol 2000;19:1–17.
7 Elston RC, Shete S: Adding further power to the Haseman and Elston's (1972) method. GeneScreen 2000;1:63–64.
8 Forrest WF, Feingold E: Composite statistics for QTL mapping with moderately discordant sibling pairs. Am J Hum Genet 2000;66:1642–1660.
9 Sham PC, Purcell S, Cherny SS, Abecasis GR: Powerful regression-based quantitative-trait linkage analysis of general pedigrees. Am J Hum Genet 2002;71:238–253.
10 Visscher PM, Hopper JL: Power of regression and maximum likelihood methods to map QTL from sib-pair and DZ twin data. Ann Hum Genet 2001;65:583–601.
11 Xu X, Weiss S, Xu X, Wei LJ: A unified Haseman-Elston method for testing linkage with quantitative traits. Am J Hum Genet 2000;67:1025–1028.
12 Sham PC, Purcell S: Equivalence between Haseman-Elston and variance-components linkage analyses for sib pairs. Am J Hum Genet 2001;68:1527–1532.
13 Amos CI, Zhu DK, Boerwinkle E: Assessing genetic linkage and association with robust components of variance approaches. Ann Hum Genet 1996;60:143–160.
14 Self SG, Liang K-Y: Asymptotic properties of maximum likelihood estimators and likelihood ratio tests under non-standard conditions. J Am Stat Assoc 1987;82:605–610.
15 Blackwelder WC, Elston RC: A comparison of sib-pair linkage tests for disease susceptibility loci. Genet Epidemiol 1985;2:85–97.
16 Allison DB, Fernandez JR, Heo M, Zhu S, Etzel C, Beasley TM, Amos CI: Bias in estimates of quantitative-trait-locus effect in genome scans: demonstration of the phenomenon and a method-of-moments procedure for reducing bias. Am J Hum Genet 2002;70:575–585.

17 Allison DB, Neale MC, Zannolli R, Schork NJ, Amos CI, Blangero J: Testing the robustness of the likelihood-ratio test in a variance-component quantitative-trait loci-mapping procedure. Am J Hum Genet 1999;65:531–544.
18 Blangero J, Williams JT, Almasy L: Robust LOD scores for variance component-based linkage analysis. Genet Epidemiol 2000;19 (suppl 1):S8–S14.
19 Amos CI, Zhu DK, Boerwinkle E: Assessing genetic linkage and association with robust components of variance approaches. Ann Hum Genet 1996;60:143–160.
20 Maclean C, Morton N, Elston R, Yee S: Skewness in commingled distributions. Biometrics 1976;32:695–699.
21 Elashoff J, Cantor R, Shain S: Power and validity of methods to identify variability genes. Genet Epidemiol 1991;8:381–388.
22 George V, Elston R: Testing the association between polymorphic markers and quantitative traits in pedigrees. Genet Epidemiol 1987; 4:193–201.
23 Go R, Elston RC, Kaplan E: Efficiency and robustness of pedigree segregation analysis. Am J Hum Genet 1978;30:28–37.
24 Demenais F, Lathrop M, Laouel J: Robustness and power of the unified model in the analysis of quantitative measurements. Am J Hum Genet 1986;38:228–234.
25 Johnson N, Kotz S: Continuous Univariate Distributions –1. Boston, Houghton Mifflin Company, 1970.
26 Vance JM, Pericak-Vance MA, Elston RC, Conneally PM, Namboodiri KK, Wappner RS, Yu PL: Evidence of genetic variation for alpha-N-acetyl-D-glucosaminidase in black and white populations: a new polymorphism. Am J Med Genet 1980;7:131–140.
27 Pericak-Vance MA, Vance JM, Elston RC, Namboodiri KK, Fogle TA: Segregation and linkage analysis of alpha-N-acetyl-D-glucosaminidase (NAG) levels in a black family. Am J Med Genet 1985;20:295–306.
28 Friedlander Y, Kark JD, Cohen T, Eisenberg S, Stein Y: Admixture analysis of high density lipoprotein cholesterol distribution in a Jerusalem population sample. Clin Genet 1983;24: 117–127.
29 Friedlander Y, Kark JD, Kidron M, Bar-On H: Univariate and bivariate admixture analyses of serum glucose and glycated hemoglobin distributions in a Jerusalem population sample. Hum Biol 1995;67:151–170.
30 Asamoah A, Wilson AF, Elston RC, Dalferes E Jr, Berenson GS: Segregation and linkage analyses of dopamine-beta-hydroxylase activity in a six-generation pedigree. Am J Med Genet 1987;27:613–621.
31 Wilson AF, Elston RC, Siervogel RM, Tran LD: Linkage of a gene regulating dopamine-beta-hydroxylase activity and the ABO blood group locus. Am J Hum Genet 1988;42:160–166.
32 Wilson AF, Elston RC, Tran LD, Siervogel RM: Use of the robust sib-pair method to screen for single-locus, multiple-locus, and pleiotropic effects: application to traits related to hypertension. Am J Hum Genet 1991;48: 862–872.
33 Amos CI, Shete S, Gu X: Variance components analysis for genetic linkage of time to onset for disease. Genet Epidemiol 2001;21(suppl 1): S768–S773.
34 Wang D, Lin S, Cheng R, Gao X, Wright FA: Transformation of sib-pair values for the Haseman-Elston method. Am J Hum Genet 2001; 68:1238–1249.

Original Paper

Human Heredity

Hum Hered 2003;55:117–124
DOI: 10.1159/000072316

Efficiency Robust Tests for Mapping Quantitative Trait Loci Using Extremely Discordant Sib Pairs

Boris Freidlin[a] Gang Zheng[c] Zhaohai Li[b,d] Joseph L. Gastwirth[b,d]

[a]Biometric Research Branch and [b]Biostatistics Branch, National Cancer Institute, [c]Office of Biostatistics Research, National Heart, Lung and Blood Institute, Bethesda, Md., and [d]Department of Statistics, George Washington University, Washington, D.C., USA

Key Words
MAX · MERT · Restricted likelihood ratio test · Robustness · Score test · Triangle constraint

Abstract
In 1972, Haseman and Elston proposed a pioneering regression method for mapping quantitative trait loci using randomly selected sib pairs. Recently, the statistical power of their method was shown to be increased when extremely discordant sib pairs are ascertained. While the precise genetic model may not be known, prior information that constrains IBD probabilities is often available. We investigate properties of tests that are robust against model uncertainty and show that the power gain from further constraining IBD probabilities is marginal. The additional linkage information contained in the trait values can be incorporated by combining the Haseman-Elston regression method and a robust allele sharing test.

Introduction

Haseman and Elston [1] proposed a regression approach for mapping quantitative trait loci (QTL) using randomly selected sib pairs. They regressed the squared difference of sib pair trait values on the estimated identical-by-descent (IBD) proportions of the sib pairs. Their method has been extended in many directions [2–9]. Sham et al. [10] developed a promising new approach that regresses the IBD sharing on the quantitative trait value. For a comprehensive review of the recent developments see Feingold [11].

Carey and Williamson [12], Eaves and Meyer [13] and Risch and Zhang [14] found that the power of testing linkage for a QTL is increased, when sib pairs with extreme trait values are used in place of the randomly selected pairs used in the original Haseman-Elston procedure. These sib pairs are called extreme discordant sib pairs (EDSP). To test whether a marker is linked to the QTL one can derive a score statistic or likelihood ratio test (LRT) using the proportion of alleles shared IBD by EDSP. Adapting the approach of Risch [15] and Holman [16], Knapp [17] found that, under some assumptions, the probabilities for the numbers of alleles IBD shared by EDSP are constrained to a triangle, and obtained a restricted likelihood test (RLRT) for linkage for QTL. Both the LRT and RLRT are efficiency-robust, i.e., they have good power properties over a family of plausible genetic models. In general, the RLRT is more powerful than LRT.

In this paper we consider a further refinement of constraints for the IBD probabilities and use efficiency robustness methods to derive several statistics that have

KARGER

Fax + 41 61 306 12 34
E-Mail karger@karger.ch
www.karger.com

0001–5652/03/0553–0117$19.50/0

Accessible online at:
www.karger.com/hhe

Boris Freidlin
Biometric Research Branch
National Cancer Institute
MSC 7334, Bethesda, MD 20892 (USA)
E-Mail freidlin@ctep.nci.nih.gov

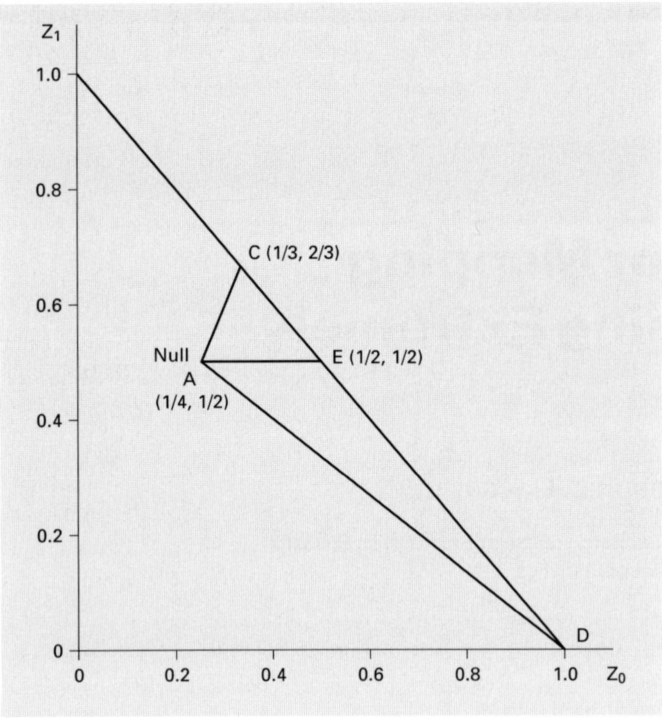

Fig. 1. Restricted probabilities for EDSP.

high power when the underlying genetic model is not precisely known. The power of these statistics can be further improved by combining them with the Haseman-Elston method as suggested by Forrest and Feingold [7].

Model and Notation

Following Hanseman and Elston [1], consider a QTL with two alleles B and b with population frequency p and $q = 1 - p$, respectively. For each individual, it is assumed that traits follow a linear model $X_{ij} = \mu + g_{ij} + e_{ij}$ ($i = 1, 2$ and $j = 1, ..., n$), where (X_{1j}, X_{2j}) are the bivariate traits for the jth sib pair, μ is the overall mean, g_{ij} is the genetic effect, and e_{ij} is the error term. As noted in Feingold [18], these residual errors incorporate polygenic, nongenetic familial as well as environmental effects, and are assumed to have a bivariate normal distribution with mean 0 and common variance. Define $g_{ij} = a$ if genotype of the ith member of the jth pair at QTL is BB, d if it is Bb, and $-a$ if it is bb. A genetic model is recessive if $d = -a$, dominant if $d = a$, and additive if $d = 0$. We also assume there is no overdominance, i.e., $a \le d \le -a$ or $-a \le d \le a$. Since the genetic effect is determined by the number of B alleles, denote $\mu_0 = \mu - a$, $\mu_1 = \mu + d$, and $\mu_2 = \mu + a$. Let $f(u, v)$ be the bivariate normal density function for the error term (e_{1j}, e_{2j}) for the jth sib pair. The correlation, ρ, between e_{1j} and e_{2j} is called residual correlation. The conditional density function for (X_{1j}, X_{2j}) can be written as $f(x_{1j} - \mu_k, x_{2j} - \mu_m)$, given that the first sib has $k = 0, 1, 2$ B alleles and the second sib has $m = 0, 1, 2$ B alleles.

Triangles for IBD Probabilties for EDSP

Consider a p_t/p_b discordant sib pair such that one sib's trait value is $\ge$ upper p_tth percentile of the distribution of the trait and the other sib's trait is $\le$ lower p_bth percentile. The population distribution function of the trait can be written as [Knapp, 17]

$$F(x) = \int_{-\infty}^{x} \{p^2h(u - \mu_2) + 2pqh(u - \mu_1) + q^2h(u - \mu_0)\}\, du,$$

where $h(u) = \int f(u, v)dv$ is the marginal distribution.

Denote by s_{km}, $0 \le k, m \le 2$, the conditional probability of a p_t/p_b discordant sib pair, given one has k alleles B and the other has m alleles B. Then, following Risch and Zhang [14] and Knapp [17], the probability that a p_t/p_b discordant sib pair shares $i = 0, 1, 2$ alleles IBD is given by

$$z_i = \text{pr}(\text{IBD} = i \mid p_t/p_b) = D_i \Big/ \sum_{J=0}^{2} D_j,$$

where

$$D_0 = \frac{p^4s_{22} + 4p^3qs_{21} + 2p^2q^2s_{20} + 4p^2q^2s_{11} + 4pq^3s_{10} + q^4s_{00}}{4}, \quad (1)$$

$$D_1 = \frac{p^3s_{22} + 2p^2qs_{21} + pqs_{11} + 2pq^2s_{10} + q^3s_{00}}{2}, \quad (2)$$

$$D_2 = \frac{p^2s_{22} + 2pqs_{11} + q^2s_{00}}{4}, \quad (3)$$

and

$$s_{km} = \int_{t-\mu_k}^{\infty} \int_{-\infty}^{b-\mu_m} f(u, v)dvdu + \int_{t-\mu_m}^{\infty} \int_{-\infty}^{b-\mu_k} f(u, v)dvdu. \quad (4)$$

Given p_t/p_b sib pairs, one tests H_0: $(z_0, z_1, z_2) = (1/4, 1/2, 1/4)$ vs. H_a: $(z_0, z_1, z_2) \neq (1/4, 1/2, 1/4)$.

When there is no overdominance, Knapp [17] showed that, for a recombination fraction $0 \le \theta \le 1/2$, the IBD probabilities z_0 and z_1 satisfy $2z_2 \le z_1$ and $z_1 \le 2z_0$. That is, in the (z_0, z_1) plane, the IBD probabilities for EDSP are constrained to a small triangle ACD (fig. 1), defined by $z_1 \le 2z_0$, $z_1 \ge 2(1 - z_0)/3$ and $z_0 + z_1 \le 1$, where the null IBD probabilities are on vertex A of the triangle.

For a recessive model ($\mu_1 = \mu_0$), Knapp [17] showed that $2s_{10} - s_{11} - s_{00} = 0$ and $s_{20} + s_{11} - s_{21} - s_{10} = 0$. Thus, from (1), (2) and (3), $D_1 - 2D_2 = p^2q(2s_{21} - s_{22} - s_{11})/2$ and $2D_0 - D_1 = p^3q(2s_{21} - s_{22} - s_{11})/2$, which imply $2D_0 - D_1 = p(D_1 - 2D_2)$, i.e.,

$$z_1 = \frac{2q}{3p+1} z_0 + \frac{2p}{3p+1}. \tag{5}$$

Equation (5) relates z_0 and z_1 for recessive model when the allele frequency p is known. When $p \to 0$, (5) goes to $z_1 = 2z_0$, which is AC, and when $p \to 1$, (5) goes to $z_1 = 1/2$, which is AE. However, for any $0 < p < 1$, when the underlying genetic model is recessive or dominant, it can be shown that the IBD probabilities, z_1 and z_0, are restricted to the triangle ACE (Appendix A). Hence, AC corresponds to a rare recessive and AE to a rare dominant. The additive model is not necessarily restricted to the triangle ACE. For case of a rare allele ($p < 0.1$) and low residual correlation numerical results indicate that the IBD probabilities (z_0, z_1) are almost always constrained to the triangle ACE. When the residual correlation is greater than 0.3 or when $p > 0.1$, (z_0, z_1) may only belong to the triangle ACD. Thus, in some common settings, it can be assumed that (z_0, z_1) are constrained to the triangle ACE.

Test Statistics

We discuss three statistics for testing whether a gene is the QTL using EDSPs. The first one is RLRT for the triangle ACE. Let n_i, $i = 0, 1, 2$, be the numbers of EDSP that share i alleles IBD and $n = n_0 + n_1 + n_2$. The likelihood function is proportional to $L(z_0, z_1) = z_0^{n_0} z_1^{n_1} z_2^{n_2} = z_0^{n_0} z_1^{n_1}(1 - z_0 - z_1)^{n_2}$. For ($z_0$, z_1) constrained in the triangle ACE, one tests H_0: $(z_0, z_1) = (1/4, 1/2)$ against H_1: $(z_0, z_1) \in \Delta ACE$ and $(z_0, z_1) \neq (1/4, 1/2)$. The RLRT can be written as

$$T_{\text{RLRT}}^{ACE} = 2 \log \left\{ \frac{\max_{(z_0, z_1) \in \Delta ACE} L(z_0, z_1)}{L(1/4, 1/2)} \right\}.$$

In contrast, the RLRT of Knapp [17], T_{RLRT}^{ACD}, tests H_0: $(z_0, z_1) = (1/4, 1/2)$ against H_1: $(z_0, z_1) \in \Delta ACD$ and $(z_0, z_1) \neq (1/4, 1/2)$. To apply T_{RLRT}^{ACE}, first, we find the restricted maximum likelihood estimators of (z_0, z_1), given in Appendix B. Then, applying Self and Liang [19] (Appendix C), we show that the asymptotic distribution of T_{RLRT}^{ACE} under the null hypothesis is a mixture of χ^2 distributions,

$$\frac{\varphi}{2\pi}\chi_2^2 + \frac{1}{2}\chi_1^2 + \left(\frac{1}{2} - \frac{\varphi}{2\pi}\right)\chi_0^2,$$

where χ_i^2 is χ^2 distribution with i degrees of freedom ($\chi_0^2 = 0$ with probability 1) and $\varphi = \cos^{-1}(6^{1/2}/3) \approx 0.1959\pi$. Thus, the mixing probabilities for χ_2^2, χ_1^2 and χ_0^2 are 0.098, 0.50 and 0.402, respectively. The asymptotic distribution of T_{RLRT}^{ACD} given in [17] is

$$\frac{\varphi_1}{2\pi}\chi_2^2 + \frac{1}{2}\chi_1^2 + \left(\frac{1}{2} - \frac{\varphi_1}{2\pi}\right)\chi_0^2,$$

where $\varphi_1 = \cos^{-1}(1/3) \approx 0.3918\pi$, so the mixing probabilities are 0.1959, 0.50 and 0.3041, respectively.

Our second statistic is a restricted score statistic. As in Whittemore and Tu [20], we reparameterize the triangle ACD. Let $G = (a, 1 - a)$ be a point on the boundary CD of the triangle ACD such that the true IBD probabilities (z_0, z_1) are on AG, where $1/3 \leq a \leq 1$. In this setting, IBD probabilities can be written as $(z_0, z_1) = r(1/4, 1/2) + (1 - r)(a, 1 - a) = (r/4 + a(1 - r), r/2 + (1 - a)(1 - r))$ for $0 \leq r \leq 1$. The null hypothesis is equivalent to H_0: $r = 0$, where a is a parameter that may depend on the genetic model, e.g., $a = 1/3$ for a rare recessive while $a = 1/2$ for a rare dominant model. For a fixed $a \in [1/3, 1]$, the score statistic is given by

$$T_a = \frac{\left(\frac{\partial}{\partial r}\log L(z_0, z_1)\right)_{r=0}}{\left\{-\text{E}\left(\frac{\partial^2}{\partial r^2}\log L(z_0, z_1)\right)_{r=0}\right\}^{1/2}} = \frac{n^{1/2}\{(n_0/n)(4a - 1) - (n_1/n)(2a - 1) - (n_2/n)\}}{(6a^2 - 4a + 1)^{1/2}}. \tag{6}$$

Note that, when the IBD probabilities are not restricted to ACD, n_i/n is the maximum likelihood estimator for z_i, $i = 0, 1, 2$. We modify (6) by replacing n_i/n by the restricted maximum likelihood estimator $\hat{z}_i$ for $i = 0, 1, 2$, which yields the restricted score statistic, $T_a^{ACD} = n^{1/2}\{\hat{z}_0(4a - 1) - \hat{z}_1(2a - 1) - \hat{z}_2\}/(6a^2 - 4a + 1)^{1/2}$, where $a \in [1/3, 1]$ and $\hat{z}_0$, $\hat{z}_1$, $\hat{z}_2$ are given in Appendix B. The restricted score statistic for the smaller triangle ACE, T_a^{ACE}, has the same expression as T_a^{ACD} but the nuisance parameter a ranges from 1/3 to 1/2.

As the score and restricted score statistics are functions of the parameter a when the underlying model generating data is unknown, they cannot be used as a is not known. Statistics based on efficiency robustness concept have been developed for this situation [21–25]. The first one is the maximin efficiency robust test (MERT) in the class of linear combinations of $\{T_a$: $1/3 \leq a \leq 1\}$. For the triangle ACD it can be shown that the MERT for this family is $T_{1/2}$. When the genetic model is in ACE the score statistics are $\{T_a$: $1/3 \leq a \leq 1/2\}$, the MERT $= (T_{1/3} + T_{1/2})/\{2(1 + \rho)\}^{1/2}$, $\rho = 6^{1/2}/3 \approx 0.816$. The second statistic for the

Table 1. Empirical power for optimal, restricted and robust statistics allele frequency $p = 0.01$, sample size 800 based on 10,000 replications

IBD prob. space	Tests	Genetic model/ascertainment percentages (p_t/p_b)								
		recessive			additive			dominant		
		10/10	0.1/10	0.05/10	10/10	10/30	1/30	10/10	10/30	1/30
Unrestricted	$T_{1/3}$	0.049	0.049	0.390	0.202	0.197	0.548	0.382	0.382	0.974
	$T_{1/2}$	0.050	0.052	0.289	0.300	0.291	0.725	0.535	0.535	0.995
	T_1	0.044	0.043	0.088	0.209	0.205	0.536	0.374	0.374	0.955
	MAX	0.047	0.047	0.317	0.249	0.241	0.632	0.457	0.457	0.988
ACD	$T_{1/3}^{ACD}$	0.048	0.060	0.520	0.311	0.305	0.675	0.512	0.512	0.989
	$T_{1/2}^{ACD}$	0.052	0.061	0.438	0.399	0.390	0.807	0.640	0.640	0.998
	T_1^{ACD}	0.047	0.051	0.145	0.304	0.297	0.660	0.501	0.501	0.977
	MAX	0.050	0.059	0.445	0.378	0.369	0.780	0.608	0.608	0.997
	T_{RLRT}^{ACD}	0.050	0.060	0.448	0.379	0.370	0.780	0.610	0.610	0.997
ACE	$T_{1/3}^{ACE}$	0.046	0.058	0.508	0.346	0.338	0.743	0.575	0.575	0.996
	$T_{1/2}^{ACE}$	0.051	0.061	0.461	0.397	0.388	0.810	0.643	0.643	0.998
	MERT	0.052	0.065	0.513	0.392	0.383	0.798	0.628	0.628	0.998
	MAX	0.051	0.064	0.505	0.391	0.383	0.798	0.629	0.629	0.998
	T_{RLRT}^{ACE}	0.050	0.062	0.502	0.384	0.374	0.789	0.618	0.618	0.998

Recessive model 10/10 $\mu_0 = -2$, $\mu_1 = -2$, $\mu_2 = 2$, $z_0 = 0.250$, $z_1 = 0.500$, $z_2 = 0.250$.
Recessive model 0.1/10 $\mu_0 = -2$, $\mu_1 = -2$, $\mu_2 = 2$, $z_0 = 0.251$, $z_1 = 0.501$, $z_2 = 0.248$.
Recessive model 0.05/10 $\mu_0 = -2$, $\mu_1 = -2$, $\mu_2 = 2$, $z_0 = 0.258$, $z_1 = 0.517$, $z_2 = 0.224$.
Additive model 10/10 $\mu_0 = -2$, $\mu_1 = 0$, $\mu_2 = 2$, $z_0 = 0.267$, $z_1 = 0.500$, $z_2 = 0.232$.
Additive model 10/30 $\mu_0 = -2$, $\mu_1 = 0$, $\mu_2 = 2$, $z_0 = 0.267$, $z_1 = 0.500$, $z_2 = 0.232$.
Additive model 1/30 $\mu_0 = -2$, $\mu_1 = 0$, $\mu_2 = 2$, $z_0 = 0.282$, $z_1 = 0.500$, $z_2 = 0.218$.
Dominant model 10/10 $\mu_0 = -2$, $\mu_1 = 2$, $\mu_2 = 2$, $z_0 = 0.275$, $z_1 = 0.500$, $z_2 = 0.224$.
Dominant model 10/30 $\mu_0 = -2$, $\mu_1 = 2$, $\mu_2 = 2$, $z_0 = 0.275$, $z_1 = 0.500$, $z_2 = 0.224$.
Dominant model 1/30 $\mu_0 = -2$, $\mu_1 = 2$, $\mu_2 = 2$, $z_0 = 0.308$, $z_1 = 0.500$, $z_2 = 0.192$.
Trinomial probabilities z were obtained using results from Knapp [17].

family $\{T_a\colon 1/3 \le a \le 1\}$ is MAX = $\max(T_{1/3}, T_1)$. When the null correlation between two extreme statistics is low to moderate, e.g., $\rho \le 0.75$, then MAX is more powerful than MERT, and when $\rho \ge 0.75$, MERT and MAX have similar power [23].

Applying MERT and MAX to restricted IBD probabilities (z_0, z_1), we replace the extreme pair ($T_{1/3}$, T_1) for the triangle *ACD* and ($T_{1/3}$, $T_{1/2}$) for the triangle *ACE* in MERT and MAX by ($T_{1/3}^{ACD}$, T_1^{ACD}) and $T_{1/3}^{ACE}$, $T_{1/2}^{ACE}$), respectively. Note that even when $p > 0.1$, and $\rho > 0.3$, the region where (z_0, z_1) may not be restricted to *ACE*, the proposed tests are still valid.

Power Comparison

To evaluate the power characteristics of the new tests under the common genetic models (recessive, additive and dominant) we conducted a simulation study of the candidate gene case for a range of population parameters and ascertainment methods. The simulations were performed as follows: first, for a given allele frequency p and ascertainment parameters p_t/p_b IBD probabilities z_0, z_1 and z_2 were calculated as in Knapp [17]. The numbers of 0, 1 and 2 sibs IBD were generated from the trinomial distribution with parameters z_0, z_1 and z_2 and sample size 800. This corresponds to an ascertainment where sib pairs are sampled and phenotyped until 800 extremely discordant pairs are selected. All tests use 0.05 two-sided significance level. Critical values for the restricted tests were obtained empirically under the null distribution of IBD probabilities. For T_{RLRT}^{ACE} the empirical critical values were quite close to the theoretical asymptotic values given in the previous section (within 3%). A SAS IML program to implement the proposed tests is available from the first author.

Table 1 presents empirical power estimates for allele frequency $p = 0.01$. Note that for rare alleles under a reces-

Table 2. Empirical power for optimal, restricted and robust statistics allele frequency $p = 0.1$, sample size 800 based on 10,000 replications

IBD prob. space	Tests	Genetic model/ascertainment percentages (p_t/p_b)								
		recessive			additive			dominant		
		10/10	5/10	2.5/10	10/10	10/30	1/30	10/10	10/30	1/30
Unrestricted	$T_{1/3}$	0.183	0.607	0.999	0.331	0.183	0.550	0.675	0.540	0.978
	$T_{1/2}$	0.156	0.494	0.993	0.484	0.274	0.742	0.824	0.692	0.995
	T_1	0.072	0.155	0.532	0.342	0.196	0.566	0.613	0.488	0.943
	MAX	0.153	0.536	0.999	0.403	0.227	0.649	0.745	0.602	0.988
ACD	$T_{1/3}^{ACD}$	0.289	0.729	0.999	0.461	0.287	0.678	0.782	0.668	0.991
	$T_{1/2}^{ACD}$	0.254	0.660	0.999	0.583	0.372	0.821	0.886	0.784	0.998
	T_{1}^{ACD}	0.120	0.238	0.779	0.461	0.228	0.685	0.730	0.614	0.971
	MAX	0.246	0.663	0.999	0.559	0.352	0.797	0.863	0.752	0.997
	T_{RLRT}^{ACD}	0.247	0.666	0.999	0.561	0.352	0.798	0.863	0.751	0.997
ACE	$T_{1/3}^{ACE}$	0.278	0.714	0.999	0.516	0.318	0.752	0.843	0.724	0.997
	$T_{1/2}^{ACE}$	0.266	0.678	0.999	0.589	0.370	0.824	0.888	0.788	0.998
	MERT	0.293	0.724	0.999	0.577	0.364	0.808	0.881	0.779	0.998
	MAX	0.288	0.717	0.999	0.576	0.363	0.809	0.881	0.777	0.998
	T_{RLRT}^{ACE}	0.288	0.715	0.999	0.568	0.355	0.802	0.876	0.769	0.998

Recessive model 10/10 $\mu_0 = -2$, $\mu_1 = -2$, $\mu_2 = 2$, $z_0 = 0.257$, $z_1 = 0.510$, $z_2 = 0.233$.
Recessive model 5/10 $\mu_0 = -2$, $\mu_1 = -2$, $\mu_2 = 2$, $z_0 = 0.264$, $z_1 = 0.520$, $z_2 = 0.216$.
Recessive model 2.5/10 $\mu_0 = -2$, $\mu_1 = -2$, $\mu_2 = 2$, $z_0 = 0.281$, $z_1 = 0.544$, $z_2 = 0.174$.
Additive model 10/10 $\mu_0 = -0.6$, $\mu_1 = 0$, $\mu_2 = 0.6$, $z_0 = 0.274$, $z_1 = 0.500$, $z_2 = 0.226$.
Additive model 10/30 $\mu_0 = -0.6$, $\mu_1 = 0$, $\mu_2 = 0.6$, $z_0 = 0.267$, $z_1 = 0.500$, $z_2 = 0.233$.
Additive model 1/30 $\mu_0 = -0.6$, $\mu_1 = 0$, $\mu_2 = 0.6$, $z_0 = 0.283$, $z_1 = 0.500$, $z_2 = 0.218$.
Dominant model 10/10 $\mu_0 = -0.4$, $\mu_1 = 0.4$, $\mu_2 = 0.4$, $z_0 = 0.285$, $z_1 = 0.502$, $z_2 = 0.213$.
Dominant model 10/30 $\mu_0 = -0.4$, $\mu_1 = 0.4$, $\mu_2 = 0.4$, $z_0 = 0.280$, $z_1 = 0.502$, $z_2 = 0.218$.
Dominant model 1/30 $\mu_0 = -0.4$, $\mu_1 = 0.4$, $\mu_2 = 0.4$, $z_0 = 0.288$, $z_1 = 0.502$, $z_2 = 0.209$.
Trinomial probabilities z were obtained using results from Knapp [17].

sive model the power of EDSP tests is very sensitive to the ascertainment parameters p_t. This happens because homozygous genotypes are rare (0.01% for $p = 0.01$) and an upper ascertainment fraction, p_t, close to p^2 has to be used to achieve meaningful power. Table 1 indicates that among tests restricted to triangle *ACD* two robust tests MAX and $T_{1/2}^{ACD}$ (*MERT*) along with the T_{RLRT}^{ACD} offer considerable improvement in power over the unrestricted tests across the three genetic models. The power gain from further restriction of the sample space is modest, with two robust tests (MERT and MAX) having power at least as high as that of T_{RLRT}^{ACD} under additive and dominant models and a marginal improvement under a recessive model. The results are similar for $p = 0.1$ and $p = 0.3$ (table 2 and 3, respectively) and when residual correlation is present (results not shown).

Recently several approaches for optimizing the power of QTL tests have been proposed. To incorporate the additional linkage information contained in the observed trait values Forest and Feingold [7] combined allele sharing and regression approaches using a linear combination of the EDSP and the Haseman-Elston statistics. To investigate the power characteristics of this combination statistic under the three genetic models and different ascertainment strategies a simulation study was conducted. We considered statistic, $T_{FFM} = \omega_1 MERT^{ACE} + \omega_2 \beta_{HE}$, a linear combination of the *MERT* (restricted to triangle *ACE*) and the β_{HE}, the Haseman-Elston regression slope estimate, with weights $(\omega_1, \omega_2) = (0.256, 0.969)$ as suggested in [7]. Each replication in the simulation was performed by repeating the following steps until the required number (800) of EDSPs with the specified degree of discordancy was ascertained: (1) parental genotypes were generated assuming Hardy-Weinberg equilibrium and random mating; (2) from the parental genotypes two sib genotypes were generated, and (3) conditionally on these sib genotypes (from step 2), sib trait values were generated according to the bivariate normal distribution model.

Table 3. Empirical power for optimal, restricted and robust statistics allele frequency $p = 0.3$, sample size 800 based on 10,000 replications

IBD prob. space	Tests	Genetic model/ascertainment percentages (p_t/p_b)								
		recessive			additive			dominant		
		10/10	10/30	1/30	10/10	10/30	1/30	10/10	10/30	1/30
Unrestricted	$T_{1/3}$	0.466	0.349	0.940	0.370	0.202	0.450	0.506	0.242	0.446
	$T_{1/2}$	0.447	0.339	0.920	0.534	0.304	0.632	0.604	0.314	0.543
	T_1	0.192	0.149	0.498	0.383	0.215	0.465	0.371	0.193	328
	MAX	0.424	0.315	0.921	0.454	0.253	0.543	0.530	0.266	0.467
ACD	$T_{1/3}^{ACD}$	0.594	0.478	0.973	0.500	0.310	0.500	0.635	0.359	0.575
	$T_{1/2}^{ACD}$	0.577	0.464	0.965	0.638	0.403	0.729	0.710	0.417	0.651
	T_1^{ACD}	0.282	0.231	0.632	0.510	0.310	0.591	0.498	0.283	0.446
	MAX	0.566	0.451	0.965	0.608	0.382	0.696	0.680	0.395	0.621
	T_{RLRT}^{ACD}	0.570	0.453	0.967	0.609	0.383	0.696	0.682	0.399	0.624
ACE	$T_{1/3}^{ACE}$	0.595	0.481	0.974	0.566	0.347	0.652	0.670	0.387	0.612
	$T_{1/2}^{ACE}$	0.593	0.478	0.970	0.638	0.401	0.730	0.720	0.422	0.662
	MERT	0.622	0.501	0.978	0.624	0.394	0.714	0.720	0.428	0.660
	MAX	0.616	0.496	0.977	0.624	0.392	0.716	0.715	0.425	0.656
	T_{RLRT}^{ACE}	0.611	0.490	0.976	0.614	0.385	0.708	0.707	0.417	0.648

Recessive model 10/10 $\mu_0 = -0.45$, $\mu_1 = -0.45$, $\mu_2 = 0.45$, $z_0 = 0.266$, $z_1 = 0.512$, $z_2 = 0.221$.
Recessive model 10/30 $\mu_0 = -0.45$, $\mu_1 = -0.45$, $\mu_2 = 0.45$, $z_0 = 0.264$, $z_1 = 0.510$, $z_2 = 0.225$.
Recessive model 1/30 $\mu_0 = -0.45$, $\mu_1 = -0.45$, $\mu_2 = 0.45$, $z_0 = 0.280$, $z_1 = 0.522$, $z_2 = 0.197$.
Additive model 10/10 $\mu_0 = -0.4$, $\mu_1 = 0$, $\mu_2 = 0.4$, $z_0 = 0.276$, $z_1 = 0.499$, $z_2 = 0.225$.
Additive model 10/30 $\mu_0 = -0.4$, $\mu_1 = 0$, $\mu_2 = 0.4$, $z_0 = 0.268$, $z_1 = 0.500$, $z_2 = 0.232$.
Additive model 1/30 $\mu_0 = -0.4$, $\mu_1 = 0$, $\mu_2 = 0.4$, $z_0 = 0.279$, $z_1 = 0.499$, $z_2 = 0.222$.
Dominant model 10/10 $\mu_0 = -0.27$, $\mu_1 = 0.27$, $\mu_2 = 0.27$, $z_0 = 0.276$, $z_1 = 0.505$, $z_2 = 0.219$.
Dominant model 10/30 $\mu_0 = -0.27$, $\mu_1 = 0.27$, $\mu_2 = 0.27$, $z_0 = 0.266$, $z_1 = 0.503$, $z_2 = 0.230$.
Dominant model 1/30 $\mu_0 = -0.27$, $\mu_1 = 0.27$, $\mu_2 = 0.27$, $z_0 = 0.274$, $z_1 = 0.505$, $z_2 = 0.222$.
Trinomial probabilities z were obtained using results from Knapp [17].

Table 4. Empirical power for MERT^{ACE} (restricted to triangle *ACE*), β_{HE} and T_{FFM} allele frequency $p = 0.1$, sample size 800 based on 10,000 replications

Tests	Genetic model/ascertainment percentage (p_t/p_b)								
	recessive			additive			dominant		
	10/10	25/25	50/50	10/10	25/25	50/50	10/10	25/25	50/50
MERT^{ACE}	0.215	0.122	0.099	0.765	0.722	0.586	0.781	0.675	0.602
β_{HE}	0.861	0.808	0.805	0.168	0.334	0.702	0.152	0.301	0.740
T_{FFM}	0.487	0.325	0.279	0.819	0.829	0.801	0.820	0.783	0.818

Table 4 presents the simulation results for $p = 0.1$ and ascertainment parameter values p_t/p_b of 10/10, 25/25 and 50/50. Combining Haseman-Elston test with the robust allele sharing test increased the power of the alleles sharing test in the situations examined. For the dominant and additive models the gain in power is modest when extremely discordant (10/10) pairs are ascertained. The power gain is noticeably greater when moderately discordant pairs are used. When the trait follows a recessive model, ED sib pairs in which one sib is homozygous (BB) with a high trait value provide most of the information. In order to capture linkage information using allele sharing methods the proportion of these homozygous individuals in the upper ascertainment region must be increased. This requires a large screening sample size. For example, if the null hypothesis is true the probability of obtaining a sib pair with one sib in the upper 1% and the other in the lower 30% is 0.003. Thus, about 66,666 sib pairs would be screened to obtain 200 EDSPs. The results indicate that employing a combination statistic should enable one to maintain a reasonable power while using a somewhat lower cut-off value to define an EDSP. Further gain may be obtained by simultaneous adjustment of the weights (ω_1, ω_2) and the ascertainment parameters p_t/p_b.

Discussion

The power of tests for linkage using sib pairs is increased when genetic knowledge constraining their IBD probabilities is incorporated [16, 17, 19]. The triangle to which these IBD probabilities are constrained in case of no over-dominance [17] is smaller for the basic genetic models when the allele frequency $p < 0.1$ and the residual correlation $\rho < 0.3$. The simulations showed that when these conditions hold the two efficiency robust tests and the restricted likelihood ratio tests have similar powers across the models studied. Relative to the tests designed for the less restrictive assumption of no over-dominance a nontrivial increase in power is only observed for the recessive model. Thus, if one is unsure whether the additional restrictions will be satisfied, the T^{ACD}_{RLRT} and robust IBD sharing tests developed for Knapp's situation (restricted to the triangle ACD) should be used, preferably in combination with the Haseman-Elston regression statistic. This is especially relevant in situations where the allele is rare ($p < 0.01$) as the sibs with the high trait value need to be in the extreme tail in order for EDSP-based test to have meaningful power.

Acknowledgments

We would like to thank the referees for their helpful suggestions. The research of Zhaohai Li was supported in part by EY14478 of The National Eye Institute and the work of Joseph Gastwirth was supported in part by SBR-9807731 of The National Science Foundation.

Appendix

A. Probabilities Constrained to the Triangle ACE

We only need to show, for any recombination fraction $0 \le \theta \le 1/2$, $z_1 \ge 1/2$. First, let $\theta = 0$. Then $z_1 \ge 1/2$ is equivalent to $D_1 \ge D_0 + D_2$. Applying (1)–(4),

$$D_1 - (D_0 + D_2) = \frac{1}{2} p^2 q^2 \left\{ \int_{t-\mu_2}^{t-\mu_1} \int_{b-\mu_2}^{b-\mu_1} dF(u, v) + \int_{t-\mu_1}^{t-\mu_0} \int_{b-\mu_1}^{b-\mu_0} dF(u, v) \right.$$

$$\left. - \int_{t-\mu_2}^{t-\mu_1} \int_{b-\mu_1}^{b-\mu_0} dF(u, v) - \int_{t-\mu_1}^{t-\mu_0} \int_{b-\mu_2}^{b-\mu_1} dF(u, v) \right\}.$$

Hence, for the recessive model ($\mu_0 = \mu_1$) or dominant model ($\mu_1 = \mu_2$), $D_1 - (D_0 + D_2) \ge 0$. Second, for $\theta > 0$, denote z_i by z_i^θ. Knapp [17] proved that $z_1^\theta \le 2z_0^\theta$. Thus, we need to show $z_1^\theta \ge 1/2$. Note that $z_1^\theta = 2\psi(1-\psi)z_0 + [\psi^2 + (1-\psi)^2]z_1 + 2\psi(1-\psi)z_2$, where $\psi = \theta^2 + (1-\theta)^2$ and $z_1 \ge 1/2$. Hence, $z_1^\theta = 2\psi(1-\psi)z_0 + [\psi^2 + (1-\psi)^2]z_1 + 2\psi(1-\psi)(1 - z_0 - z_1) = [\psi - (1-\psi)]^2 z_1 + 2\psi(1-\psi) \ge (2\psi - 1)^2/2 + 2\psi(1-\psi) = 1/2$, i.e., (z_0^θ, z_1^θ) are constrained to the triangle ACE.

B. Restricted Maximum Likelihood Estimator for the Triangle ACE

Denote the restricted maximum likelihood estimator (MLE) by $(\hat{z}_0, \hat{z}_1)$. First, find unrestricted MLE. If it is contained in the triangle ACE, ten it is also the restricted MLE. Otherwise, the restricted MLE should be on one of the three boundaries of the triangle ACE. The likelihood function, evaluated at the boundary CE, is 0. Thus, we only consider the boundaries: AC where $z_1 = 2z_0$ and AE where $z_1 = 1/2$. Note that $\max_{(z_0, z_1) \in AC} L(z_0, z_1) = \max_{1/4 \le z_0 \le 1/3} L(z_0, 2z_0)$, from which $\hat{z}_0 = (n_0 + n_1)/(3n)$ and $\hat{z}_1 = 2\hat{z}_0$ when $n_0 + n_1 \ge 3n/4$ and $(\hat{z}_0, \hat{z}_1) = (1/4, 1/2)$ when $n_0 + n_1 \le 3n/4$. On the boundary AE, $\max_{(z_0, z_1) \in AE} L(z_0, z_1) = \max_{1/4 \le z_0 \le 1/2} L(z_0, 1/2)$ from which $\hat{z}_0 = n_0/(2(n - n_1))$ when $2n_0 + n_1 \ge n$ and $\hat{z}_0 = 1/4$ when $2n_0 + n_1 \le n$. Since $(\hat{z}_0, \hat{z}_1)$ maximizes $L(z_0, z_1)$ on both AC and AE, combining the results yields

$$(\hat{z}_0, \hat{z}_1) = \left(\frac{n_0 + n_1}{3n}, \frac{2(n_0 + n_1)}{3n} \right),$$

if $n_0 + n_1 \ge 3n/4$ and $2n_0 + n_1 \le n$;

$$= (1/4, 1/2), \text{ if } n_0 + n_1 \le 3n/4 \text{ and } 2n_0 + n_1 \le n;$$

$$= \left(\frac{n_0}{2(n - n_1)}, 1/2 \right), \text{ if } n_0 + n_1 \le 3n/4 \text{ and } 2n_0 + n_1 \ge n;$$

$$= \text{either} \left(\frac{n_0 + n_1}{3n}, \frac{2(n_0 + n_1)}{3n} \right) \text{ or } \left(\frac{n_0}{2(n - n_1)}, 1/2 \right), \quad (7)$$

otherwise.

Given the data (n_0, n_1, n_2), evaluating $L(z_0, z_1)$ at both points in (7) yields the restricted MLE $(\hat{z}_0, \hat{z}_1)$, the point that maximizes the likelihood function.

C. Asymptotic Null Distribution of RLRT for the Triangle ACE

For testing H_0: $(z_0, z_1) = (1/4, 1/2)$ against H_1: $(z_0, z_1) \in \Delta ACE - (1/4, 1/2)$, both parameters of interest (z_0, z_1) are on the boundary of the triangle ACE. From case 7 of Self and Liang [19], the asymptotic distribution of RLRT follows a mixture of χ_0^2, χ_1^2 and χ_2^2 with mixing probabilities $1/2 - \varphi/(2\pi)$, $1/2$ and $\varphi/(2\pi)$, respectively. The factor φ

depends on the Fisher information matrix under the null hypothesis, $I = (I_{ij})_{2 \times 2}$, where $I_{ij} = -E(\partial^2 L(z_0, z_1)/(\partial z_i \partial z_j)$ for $i, j = 0, 1$. These values are $I_{00} = 8$, $I_{01} = 4$ and $I_{11} = 6$.

Let $I = P\lambda P'$ be a spectral decomposition of I. Then, by a linear transformation of IBD probabilities (z_0, z_1) on boundaries AC and AE: $\lambda^{1/2}P'(AC)'$ and $\lambda^{1/2}P'(AE)'$, we can apply the result of Self and Liang to obtain φ, given by

$$\cos\varphi = \frac{(AC)I(AE)'}{\|\lambda^{1/2}P'(AC)'\| \, \|\lambda^{1/2}P'(AE)'\|},$$

where $(AC) = (1/3, 2/3) - (1/4, 1/2) = (1/12, 1/6)$ and $(AE) = (1/2, 1/2) - (1/4, 1/2) = (1/4, 0)$. Calculations show that $(AC)I(AE)' = I_{00}/48 + I_{01}/24$, $\|\lambda^{1/2}P'(AC)'\|^2 = (I_{00} + 4I_{11} + 4I_{01})/144$, and $\|\lambda^{1/2} P'(AE)'\|^2 = I_{00}/16$, yielding $\cos\varphi = 6^{1/2}/3$.

References

1 Haseman JK, Elston RC: The investigation of linkage between a quantitative trait and a marker locus. Behav Genet 1972;2:3–19.

2 Elston RC, Buxbaum S, Jacobs KB, Olson JM: Haseman and Elston revisited. Genet Epidemiol 2000;19:1–17.

3 Xu X, Weiss S, Xu X, Wie LJ: A unified Haseman-Elston method for testing linkage with quantitative traits. Am J Hum Genet 2000;67:1025–1028.

4 Ziegler A: The new Haseman-Elston method and the weighted pairwise correlation statistic are variations on the same theme. Biometrical J 2001;43:697–702.

5 Fernandez JR, Etzel C, Beasley TM, Shete S, Amos CI, Allison DB: Improving the power of sib pair quantitative trait loci detection by phenotype winsorization. Hum Hered 2002;53:59–67.

6 Drigalenko E: How sib pairs reveal linkage. Am J Hum Genet 1998;63:1442–1245.

7 Forrest W, Feingold E: Composite statistics for QTL mapping with moderately discordant sibling pairs. Am J Hum Genet 2000;66:1642–1660.

8 Forrest W: Weighting improves the 'New Haseman-Elston' method. Hum Hered 2001;52:47–54.

9 Sham PC, Parcell S: Equivalence between Haseman-Elston and variance-components linkage analyzes for sib pairs. Am J Hum Genet 2001;68:1527–1532.

10 Sham PC, Purcell S, Cherny SS, Abecasis GR: Powerful regression-based quantitative-trait linkage analysis of general pedigrees. Am J Hum Genet 2002;71:238–253.

11 Feingold E: Regression-based quantitative-trait-locus mapping in the 21st century. Am J Hum Genet 2002;71:217–222.

12 Carey G, Williamson J: Linkage analysis of quantitative traits: Increased power by using selected samples. Am J Hum Genet 1991;49:786–796.

13 Eaves L, Meyer J: Locating human quantitative trait loci: Guidelines for the selection of sibling pairs for genotyping. Behav Genet 1994;24:443–455.

14 Risch H, Zhang H: Extreme discordant sib pairs for mapping quantitative trait loci in humans. Science 1995;268:1584–1589.

15 Risch N: Linkage strategies for genetically complex traits III. The effect of marker polymorphism on analysis of affected relative pairs. Am J Hum Genet 1990;46:242–253.

16 Holmans P: Asymptotic properties of affected-sib-pair linkage analysis. Am J Hum Genet 1993;52:362–374.

17 Knapp M: Evaluation of a restricted likelihood ratio test for mapping quantitative trait loci with extreme discordant sib pairs. Ann Hum Genet 1998;62:75–78.

18 Feingold E: Methods for linkage analysis of quantitative trait loci in humans. Theor Popul Biol 2001;60:167–180.

19 Self SG, Liang KY: Asymptotic properties of maximum likelihood estimators and likelihood ratio tests under nonstandard conditions. J Am Stat Ass 1987;82:605–610.

20 Whittemore AS, Tu I-P: Simple, robust linkage tests for affected sibs. Am J Hum Genet 1998;62:1228–1242.

21 Gastwirth JL: On robust procedures. J Am Stat Ass 1966;61:929–948.

22 Gastwirth JL: The use of maximin efficiency robust tests in combining contingency tables and survival analysis. J Am Stat Ass 1985;80:380–384.

23 Freidlin B, Podgor MJ, Gastwirth JL: Efficiency robust tests for survival or ordered categorical data. Biometrics 1999;55:883–886.

24 Shih M, Whittemore AS: Allele-sharing among affected relatives: Nonparametric methods for identifying genes. Stat Methods Med Res 2001;10:27–55.

25 Gastwirth JL, Freidlin B: On power and efficiency robust linkage tests for affected sibs. Ann Hum Genet 2000;64:443–453.

Original Paper

Human Heredity

Hum Hered 2003;55:125–131
DOI: 10.1159/000072317

Regression-Based Sib Pair Linkage Analysis for Binary Traits

Maurice P.A. Zeegers[a,b] John P. Rice[c] Frühling V. Rijsdijk[b]
Goncalo R. Abecasis[d] Pak C. Sham[b,e]

[a]Department of Epidemiology, Faculty of Health Science, Maastricht University, Maastricht, The Netherlands; [b]Social, Genetic and Developmental Psychiatry Centre, Institute of Psychiatry, King's College London, London, UK; [c]School of Medicine, Washington University, St. Louis, Mo., [d]Department of Biostatistics, School of Public Health, University of Michigan, Ann Arbor, Mich., USA; [e]Department of Psychiatry, University of Hong Kong, Queen Mary Hospital, Hong Kong, China

Key Words

Binary traits · Linkage · Haseman-Elston · Variance components

Abstract

The Haseman-Elston (HE) regression method offers a mathematically and computationally simpler alternative to variance-components (VC) models for the linkage analysis of quantitative traits. However, current versions of HE regression and VC models are not optimised for binary traits. Here, we present a modified HE regression and a liability-threshold VC model for binary-traits. The new HE method is based on the regression of a linear combination of the trait squares and the trait cross-product on the proportion of alleles identical by descent (IBD) at the putative locus, for sibling pairs. We have implemented both the new HE regression-based method and have performed analytic and simulation studies to assess its type 1 error rate and power under a range of conditions. These studies showed that the new HE method is well-behaved under the null hypothesis in large samples, is more powerful than both the original and the revisited HE methods, and is approximately equivalent in power to the liability-threshold VC model.

Introduction

The original Haseman-Elston (HE) method, based on the regression of squared trait difference on the estimated proportion of alleles identical by descent (IBD) of sibling pairs, was the first proposed method for non-parametric linkage analysis of quantitative traits [Haseman and Elston, 1972]. Subsequent modifications of the method to improve its statistical power have also focused on quantitative phenotypes [Drigalenko 1998; Elston et al., 2000; Forrest 2001; Sham and Purcell 2001; Visscher and Hopper 2001; Wright 1997; Xu et al., 2000]. The method has been applied to binary traits by simply assigning the values 1 and 0 for the presence and absence of the trait and proceeding to treat this binary variable as if it were continuous [Dawson et al., 1990; S.A.G.E. 4.0, 1999; Elston et al., 2000]. Unlike popular affected sib pair methods [Suarez et al., 1978; Risch, 1990; Holmans, 1993], this regression method can use information from both affected and unaffected individuals. In this paper, we adopt the regression approach for binary traits, but follow the procedures of Sham and Purcell [2001] to derive a regression method that optimally combines information from the squares and cross-products of the mean-centred sibling trait values. The proposed method involves regress-

KARGER

Fax + 41 61 306 12 34
E-Mail karger@karger.ch
www.karger.com

0001–5652/03/0553–0125$19.50/0

Accessible online at:
www.karger.com/hhe

Dr. P.C. Sham
Social, Genetic and Developmental Psychiatry Centre
Institute of Psychiatry, King's College London
PO Box 080, London SE5 8AF (UK)
Tel. +44 207 848 0892, Fax +44 207 848 0866, E-Mail p.sham@iop.kcl.ac.uk

ing this optimal linear combination on the estimated proportion of alleles IBD, of sibling pairs.

The variance components model for partitioning phenotypic variance into genetic and environmental components was initially developed for quantitative traits [Fisher, 1918] but has been extended to complex presence-absence traits through the introduction of an underlying continuous liability with a discrete threshold [Crittenden, 1961]. Under this liability-threshold model, concordance data in twins can be converted to tetrachoric correlations and then to heritability estimates of the liability to disease [Falconer, 1965]. Just as the variance components model can be used for QTL linkage analysis, so can its liability-threshold extension. Since this method is based on a model that is commonly believed to be appropriate for complex diseases, we have developed it to provide a standard to which the new HE method can be compared.

In this article, we describe the details of the new HE regression and the liability-threshold VC methods, and conduct analytic studies to show that they have similar power to detect a QTL. We also report the results of simulation studies that compare the power of the new HE method to those of existing HE methods, and assess the robustness of the new method to phenotypically ascertained samples such as affected sib pairs.

Method

Haseman-Elston Regression for Binary Traits

The affection status of an individual for a disease is coded as 0 (absent) and 1 (present). Let the affection status of two siblings be denoted as X and Y. The disease prevalence is $K = P(X = 1) = P(Y = 1)$, and the sibling relative risk is $\lambda = [P(Y = 1 \mid X = 1)]/[P(Y = 1)$. Sib pairs can be classified into three classes: concordant affected, with probability $K^2\lambda$; discordant, with probability $2(K - K^2\lambda)$); and concordant unaffected, with probability $1-2K + K^2\lambda$. Based on these probabilities, the mean, variance and covariance of any function of X and Y can be calculated. In Sham and Purcell [2001], a linear combination of the squared difference and squared sum was regressed on the estimated proportion of alleles IBD ($\hat{\pi}$). The quantity $\hat{\pi}$ is the independent variable, which needs to be estimated from the marker genotype data of the sibling pair using programs such as Genehunter or Merlin [Kruglyak et al., 1996; Abecasis et al., 2002]. A linear combination of squared differences and squared sums is equivalent to a linear combination of squares and products [Sham et al., 2002]. We therefore seek an optimal linear combination of X^2, Y^2 and XY that would form the dependent variable, to be regressed on $\hat{\pi}$. Writing $(X^2, Y^2, XY)^T$ as the vector $\mathbf{u}$, then the theoretical covariance matrix of $\mathbf{u}$ in the population is:

$$\Sigma = K \otimes \begin{bmatrix} (1-K) & K(\lambda-1) & K\lambda(1-K) \\ K(\lambda-1) & (1-K) & K\lambda(1-K) \\ K\lambda(1-K) & K\lambda(1-K) & K\lambda(1-K^2\lambda) \end{bmatrix}$$

The Kronecker product $\otimes$ indicates that each element of the 3×3 matrix is multiplied by the scalar K. The inverse of this covariance matrix is:

$$\Sigma^{-1} = [K(1-2K+\lambda K^2)]^{-1} \otimes \begin{bmatrix} \frac{1-K}{1-\lambda K} & K & \frac{-(1-K)}{1-\lambda K} \\ K & \frac{1-K}{1-\lambda K} & \frac{-(1-K)}{1-\lambda K} \\ \frac{-(1-K)}{1-\lambda K} & \frac{-(1-K)}{1-\lambda K} & \frac{1-2K+\lambda K}{\lambda K(1-\lambda K)} \end{bmatrix}$$

Let Q represent the difference in covariance in the (0, 1) trait between sib pairs that share 0 allele IBD and those that share 2 alleles IBD at the quantitative trait locus (QTL), then Q is also the regression coefficient of XY on π in the population. Thus the covariance between XY and π is $Q\mathrm{Var}(\pi)$, while X^2 and Y^2 are uncorrelated with π. The population covariance matrix between π and $\mathbf{u}$ is therefore

$$\mathbf{c} = \begin{bmatrix} 0 \\ 0 \\ Q\mathrm{Var}(\pi) \end{bmatrix}$$

The population regression coefficients of π on $\mathbf{u}$ are then given by $\mathbf{c}^T\Sigma^{-1}$. After algebraic simplification, it can be shown that the predicted values of this linear regression model $\mathbf{c}^T\Sigma^{-1}\mathbf{u}$, can be rewritten in the form of the product FZ, where F and Z are the scalars:

$$F = \frac{-Q\,Var(\pi)(1-K)}{K(1-K\lambda)(1-2K+K^2\lambda)}$$

$$Z = X^2 + Y^2 - \frac{1-2K+K\lambda}{K\lambda(1-K)}XY$$

By standard regression theory, Z is the linear combination of X^2, Y^2 and XY that is maximally correlated with π, so that a regression of Z on π is expected to be a powerful test for linkage. The regression coefficient estimate obtained from a particular sample will be an estimate of F.

For selected samples we mean-centre both Z and π before the regression and constrain the constant term in the regression to 0 [Sham et al., 2001]. The expected value of π is simply 1/2, while that of Z can be obtained by recognising that the expected values of X^2 and Y^2 are K, while that of XY is $K^2\lambda$.

Liability-Threshold Variance Components Model

The liability-threshold model for QTL linkage assumes that the joint liability distribution for a sib pair, conditional on IBD sharing proportion π, is bivariate normal with correlation $0.5h^2 + (\pi - 0.5)q^2$. The expected counts of the three types of sib pairs, concordant affected, concordant unaffected, and discordant, for the three IBD classes can be obtained by numerical integration of the bivariate normal density function, given the values of the parameters h^2 (the overall heritability of liability), q^2 (QTL-specific heritability), and the liability threshold T. The likelihood of a given sib pair is given by the integral of the appropriate quadrant (determined by the trait values of the sib pair) of the bivariate normal distribution with the appropriate correlation (determined by the IBD sharing of the sib pair). The overall likelihood function of a sample is the product of the probabilities of these sib pair likelihood contributions, over all the sib pairs in the sample. The parameters h^2, q^2, and T can therefore be estimated by maximising the likelihood function. The null hypothesis that $q^2 = 0$ can be tested by a likelihood ratio test comparing a model where all three parameters are estimated to one in which q^2 is fixed to 0. We have implemented these procedures using the Mx software [Neale, 2000].

For selected samples, the VC method has to incorporate an ascertainment correction. This correction involves dividing the likelihood of the frequency data by the expected proportion remaining after omission of unobserved cells, while fixing the threshold to the z value that gives the correct population prevalence (e.g. 2.33 for a trait with prevalence 0.01, and 1.28 for trait for a trait with prevalence 0.1). For the affected sib pair design, it is necessary to 'impute' the number of sib pairs in each IBD category, who are not concordant affected and therefore not ascertained into the sample. Let the number of affected sib pairs with 0, 1 and 2 alleles IBD be n_0, n_1 and n_2, and the total number of affected sib pairs be $n = n_0 + n_1 + n_2$, then an estimate for the total number of sib pairs in the population from which the affected sib pairs were drawn is $N = n/(K^2\lambda)$. We therefore impute the numbers $0.25N - n_0$, $0.5N - n_1$, and $0.25N - n_2$ as the numbers of sib pairs in the IBD categories of 0, 1 and 2, who are either concordant unaffected or discordant and were therefore not ascertained. The cell probabilities of each of these three 'composite missing cells' were obtained by summing the cell probabilities of the corresponding missing cells. The data therefore consist of 3 observed and 3 imputed cell counts. A test of QTL effect is obtained by the difference in log-likelihood between a model in which both h^2 and q^2 are free (although these are bounded to have positive values) and one that fixes the q^2 to zero. This likelihood ratio test statistic is asymptotically distributed under the null hypothesis as a 50:50 mixture of a point mass at 0 and a χ_1^2 random variable.

Mean-IBD Method

This method is applicable only to the affected sib pair design and simply compares the sum of the IBD sharing proportions in the affected sib pairs of the sample against its expectation and variance under the null hypothesis. The χ^2 statistic with one degree of freedom is:

$$\chi_1^2 = \frac{\left(\Sigma\hat{\pi} - n/2\right)^2}{n/8}$$

where n is the number of affected sib pairs. This test is one-sided: only excess allele sharing is considered to be evidence for linkage.

Simulation and Analytic Studies

Although the above methods are non-parametric in the sense that they do not assume an underlying trait model, for the purpose of evaluating their properties we have adopted a liability-threshold model. In this model, the underlying liability is determined by additive QTL effect, polygenic background and non-shared environmental influences. The QTL is assumed to be biallelic; the trait-increasing allele being labelled B, with allele frequency p. The parameters of the model are: disease prevalence K, sibling relative risk λ, the overall heritability of liability h^2, and the QTL-specific heritability q^2. The three QTL genotypes are assumed to have liability distributions that are univariate normal with different means but equal variance.

Several sets of model parameters were used to generate expected counts and simulated data for further analyses. The QTL-specific heritability, q^2, was set at 0, 20, 40 or 60%. Two alternative sets of values for K, λ and h^2 were used: trait 1 is characterised by $K = 0.01$, $\lambda = 10$, and $h^2 = 0.80$, considered to be representative of a relatively rare complex disorder such as schizophrenia [Cardno et al., 1999; Egan et al., 2001]; while trait 2 is characterised by $K = 0.10$, $\lambda = 2.5$ and $h^2 = 0.60$, considered to be representative of common diseases such as prostate cancer [Neider et al., 2002; Page et al., 1997; Zeegers et al., 2002] and coronary heart disease [Snieder, 2000] in an older population.

For each genetic model, we calculated the expected proportions of sib pairs with different possible combinations of trait and IBD values. Sib pairs can be classified into three IBD classes ($\pi = 0, 0.5, 1$) each having a characteristic distribution of genotypes, corresponding to those for unrelated pairs, parent-offspring pairs, and monozygotic twin pairs, respectively [Risch et al., 1990]. Given the QTL allele frequency p and the QTL-specific heritability q^2, the QTL allelic effect is given by $a = q/\sqrt{2p(1-p)}$. The liability threshold for development of disease T is obtained by equating the prevalence K to the weighted sum of the three normal integrals (for the probabilities that liability will exceed T), one for each of the three QTL genotypes. Similarly, the allele frequency p is found by equating the sibling relative risk (λ) to the weighted sum of the nine bivariate normal integrals (for the probabilities that both sibling liabilities will exceed T, one for each of the nine possible pairs of sibling genotypes), divided by K^2. Finally, sibling relative risks for sib pairs who share 0, 1 or 2 alleles IBD at the QTL (corresponding to $\pi = 0, 0.5$ and 1, respectively) can be calculated as weighted sums of the nine bivariate normal integrals divided by K^2, where the weights are IBD-specific genotyping probabilities. Based on the IBD-specific sibling relative risks, the expected counts of concordant unaffected (0,0) pairs, discordant (0,1) pairs, discordant (1,0) pairs, and concordant affected pairs were calculated for 25,000, 50,000 and 25,000 sib pairs sharing 0, 1 and 2 alleles IBD, respectively. The expected cell counts for each IBD class characterised by IBD-specific sibling relative risk λ_1, are proportional to $P_{00} = 1 - 2K + K^2\lambda_1$, $P_{01} = P_{10} = K - K^2\lambda_1$, $P_{11} = K^2\lambda_1$, respectively.

Simulation of sib pair data was performed using the statistical software package Stata [StataCorp, 2001]. Two latent variables (the residual additive genetic components and the environmental components) were generated for each sib as standard normal random variates. The residual additive genetic components were simulated to have correlation 0.5 between siblings. The latent variables representing QTL effects were simulated by first generating two independent parental diallelic genotypes assuming a gene frequency p for the trait-increasing allele B. The transmissions of these alleles to two siblings were generated keeping track of parental ancestry, enabling sibling pairs to be classified as sharing 0, 1 or 2 alleles IBD. In effect, this assumes that perfect IBD information is available. We standardised the QTL effect, initially coded as –1, 0 and 1 for the genotypes bb, Bb and BB, by subtracting the mean $2p - 1$ and dividing this by the standard deviation $\sqrt{2p(1-p)}$ [Falconer and Mackay, 1996]. The trait liabilities for each sibling were specified to be a linear combination of the residual additive genetic components, the environmental components and the QTL components assuming an overall heritability of liability (h^2) and a QTL-specific heritability of liability (q^2). The trait liabilities were subsequently dichotomised to binary (0,1) trait values according to a threshold T that gives the required population risk K. In summary, our simulation program takes values for the parameters K, h^2, q^2 and p as input and generates for each sib pair two binary trait values (X and Y) and an IBD sharing proportion π. In each data set, we simulated phenotypes for 100,000 sib pairs. Such a large number is often necessary in order to obtain a sufficient number of sib pairs with one or two affected members. For Trait 1, the simulations typically obtained fewer than 2000 sib pairs with at least one affected member, and around 100 sib pairs where both are affected. When conducting simulations under both the null and alternative hypothesis we examined 100 replicate data sets.

Apart from simple population samples (as generated above) we considered two different schemes of phenotypic selection. The first is complete ascertainment of affected individuals, in which all sib pairs with one or two affected members are included. The second is the affected sib pair (ASP) design, in which only concordant affected sib pairs are included. These two schemes were used to select sib pairs from the population samples generated above, under both schemes only a small proportion of the 100,000 sib pairs are selected from each replicate.

The new regression method was applied to data simulated under the null hypothesis, to check that the test statistics comply with their theoretical sampling distributions. In addition, we also assessed the null distributions of the original HE method [Haseman and Elston, 1972], that is a regression of $(X - Y)^2$ on π, and the revisted HE method [Elston et al., 2002], a regression of $(X - K)(Y - K)$ on π, as well as the mean-IBD test (for the ASP designs). All these methods were also applied to data simulated under a range of alternative hypotheses in order to assess their relative power. All these methods and the liability-threshold VC method were also applied to data that simply represent the expected counts under the various alternative hypotheses. This eliminates the effect of sampling variability and provides accurate estimates of the non-centrality parameters (NCP) of the tests.

For comparability we used likelihood-ratio statistics for all the above tests (except the mean-IBD test). For regression analyses, the likelihood ratio is test is equal to $-N \ln(1 - \hat{R}^2)$, where N is the number of sib pairs in the sample and $\hat{R}^2$ is the estimated proportion of variance explained by the regression [Sham and Purcell, 2001]. For all regression methods the χ_1^2 statistic was set to zero when the regression coefficient is in the opposite direction to that predicted by linkage.

Results

Distribution under the Null Hypothesis

Under the null hypothesis the asymptotic distribution of all the test statistics is a 50:50 mixture of 0 and χ_1^2. The first, second and fourth moments of this distribution are 0.5, 1.5, and 52.25, respectively [Stuart and Ord, 1994]. With 100 replicates, the mean test statistic is therefore expected to have mean 0.5 (with standard error $\sqrt{1.5 - 0.5^2}/10 = 0.118$) and standard deviation $\sqrt{1.5 - 0.5^2} = 1.18$ (with standard error $\sqrt{52.25 - 1.5^2}/10 = 0.709$). All regression methods and the mean-IBD test give χ^2 statistics with means and standard deviations well within two standard errors of their expected values (tables 1–3, condition $q^2 = 0$). This suggests that all the tests follow their theoretical distributions under the null hypothesis, and that asymptotic p values are appropriate for the sample sizes used in the simulations.

Power Comparisons

The results obtained from expected counts were very close to the average results from simulated data. The average χ^2 statistics of the proposed regression for binary traits consistently exceeded those of both the original and the revisited Haseman-Elston regression methods, and are similar to those of the variance components method in unselected and in complete ascertainment samples (tables 1–3). In complete ascertainment samples, the average χ^2 statistics of the new method even exceeded those of variance components somewhat when the QTL-specific heritability was high. Although the average χ^2 statistics of the revisited Haseman-Elston regression were higher than those of the original method in unselected samples, these two methods appeared to be identical in complete ascertainment samples (table 2). This is to be expected because the dependent variables of the two methods are linearly related, each having only two possible values (one value for discordant sib pairs and one value for concordant affected sib pairs). For the affected sib-pair design, the average χ^2 statistics of the new regression method were slightly higher than those of mean-IBD χ^2 test, and close to those of the variance components method (table 3).

Discussion

We have proposed a regression-based linkage analysis method for binary traits and shown, through simulated data and expected counts, that this method behaves well in large samples under the null hypothesis, and has greater power than both the original and the revisited HE methods under a range of alternative hypotheses. The proposed regression appears to have power similar to a more sophisticated variance components approach that uses a liability-threshold model.

The original and revisited Haseman-Elston regression methods are currently in use for linkage analysis of binary traits. They are, for example, implemented in the computer program SIBPAL2 of the SAGE package [S.A.G.E. 4.0, 1999]. The results of this study, however, suggest that it may be preferable to use the proposed binary regression method.

The new regression method can be regarded as a generalisation of the original and revisited Haseman-Elston regression methods. The dependent variable of the original HE regression analysis can be written as $X + Y - 2XY$ (since $X = X^2$ for a binary trait), while that of the revisited HE method can be written as $XY - KX - KY + K^2$, which can be transformed to $X + Y - (1/K)XY$ without affecting

Table 1. Average χ^2 statistics of original (HE-SD) and revisited (HE-CP) Haseman-Elston regression, proposed binary Haseman-Elston regression (HE-B) and variance components methods (VC) under various null and alternative hypotheses in unselected samples

Hypothesis	Expected counts				Simulated data					
	HE-SD	HE-CP	HE-B	VC	HE-SD		HE-CP		HE-B	
					mean[1]	SD	mean[1]	SD	mean[1]	SD
Null										
$q^2 = 0$										
Trait 1[2]	0.00	0.00	0.00	0.00	0.35	0.65	0.50	1.27	0.52	1.25
Trait 2[3]	0.00	0.00	0.00	0.00	0.47	1.51	0.43	0.79	0.38	0.81
Alternative										
$q^2 = 0.2$										
Trait 1[2]	4.34	20.19	23.70	24.30	5.42	5.19	19.85	8.68	23.44	10.42
Trait 2[3]	132.86	242.47	283.74	289.70	130.71	20.17	238.14	30.06	282.31	32.86
$q^2 = 0.4$										
Trait 1[2]	9.58	44.02	51.78	54.40	9.83	5.67	44.21	11.88	51.39	12.32
Trait 2[3]	272.83	498.44	583.29	602.80	275.23	34.15	498.05	42.78	591.99	46.82
$q^2 = 0.6$										
Trait 1[2]	15.21	70.70	83.00	90.10	16.20	7.90	70.96	14.01	83.14	15.43
Trait 2[3]	512.94	937.61	1,097.86	1,164.90	511.79	47.75	940.85	58.90	1,114.87	68.34

1 Average χ^2 statistics from the simulated data were obtained from 100 replicates of 100,000 sib pairs.
2 The parameters for trait 1 are disease prevalence = 0.01, sibling relative risk = 10.00, and overall heritability = 0.80.
3 The parameters for trait 2 are disease prevalence = 0.10, sibling relative risk = 2.50, and overall heritability = 0.60.

Table 2. Average χ^2 statistics of original (HE-SD) and revisited (HE-CP) Haseman-Elston regression, mean adjusted binary Haseman-Elson regression (HE-B) and variance components methods (VC) under various null and alternative hypotheses in complete ascertainment samples

Hypothesis	Expected counts				Simulated data					
	HE-SD	HE-CP	HE-B	VC	HE-SD		HE-CP		HE-B	
					mean[1]	SD	mean[1]	SD	mean[1]	SD
Null										
$q^2 = 0$										
Trait 1[2]	0.00	0.00	0.00	0.00	0.51	1.26	0.51	1.26	0.52	1.23
Trait 2[3]	0.00	0.00	0.00	0.00	0.42	0.76	0.42	0.76	0.38	0.78
Alternative										
$q^2 = 0.2$										
Trait 1[2]	22.90	22.90	23.92	23.20	22.60	9.70	22.60	9.70	23.63	10.40
Trait 2[3]	260.35	260.35	280.11	260.50	255.82	31.61	255.82	31.61	279.16	33.57
$q^2 = 0.4$										
Trait 1[2]	50.75	50.75	53.01	51.90	50.69	12.75	50.69	12.75	52.65	12.79
Trait 2[3]	541.87	541.87	582.42	541.90	541.56	45.46	541.56	45.46	592.63	47.33
$q^2 = 0.6$										
Trait 1[2]	82.68	82.68	86.29	85.90	83.40	16.26	83.40	16.26	86.89	16.60
Trait 2[3]	1,047.99	1,047.99	1,124.75	1,044.20	1,050.73	65.12	1,050.73	65.12	1,142.96	70.63

1 Average χ^2 statistics from the simulated data were obtained from 100 replicates of 100,000 sib pairs, selecting sib pairs with at least one affected members.
2 The parameters for trait 1 are disease prevalence = 0.01, sibling relative risk = 10.00, and overall heritability = 0.80.
3 The parameters for trait 2 are disease prevalence = 0.10, sibling relative risk = 2.50, and overall heritability = 0.60.

Table 3. Average χ^2 statistics of mean IBD test (M-IBD), mean adjusted binary Haseman-Elson regression (HE-B) and variance components methods (VC) under various null and alternative hypotheses in affected sib pair samples

Hypothesis	Expected counts			Simulated data			
	M-IBD	HE-B	VC	M-IBD		HE-B	
				mean[1]	SD	mean[1]	SD
Null							
$q^2 = 0$							
Trait 1[2]	0.00	0.00	0.00	0.49	1.26	0.49	1.28
Trait 2[3]	0.00	0.00	0.00	0.47	0.91	0.47	0.92
Alternative							
$q^2 = 0.2$							
Trait 1[2]	19.41	20.03	19.80	19.12	8.39	19.74	8.77
Trait 2[3]	169.28	174.70	176.20	166.77	23.91	171.88	25.05
$q^2 = 0.4$							
Trait 1[2]	42.32	43.49	44.00	42.56	11.65	44.20	12.49
Trait 2[3]	347.56	361.29	364.00	346.53	36.30	358.49	38.30
$q^2 = 0.6$							
Trait 1[2]	67.96	71.61	72.20	68.19	13.80	71.80	16.22
Trait 2[3]	652.33	661.46	680.60	657.41	46.59	667.46	49.30

1 Average χ^2 statistics from the simulated data were obtained from 100 replicates of 100,000 sib pairs, selecting affected sib pairs.
2 The parameters for trait 1 are disease prevalence = 0.01, sibling relative risk = 10.00, and overall heritability = 0.80.
3 The parameters for trait 2 are disease prevalence = 0.10, sibling relative risk = 2.50, and overall heritability = 0.60.

the significance test on the regression coefficient. The dependent variable of the proposed binary regression method can be expressed as $X + Y - \alpha XY$, where $\alpha = (1 - 2K + K\lambda)/(K\lambda(1 - K))$. It is easy to show that the new method is equivalent to the original HE when $K\lambda = 1$, and is equivalent to the revisited HE when either $K = 0.5$ or $\lambda = 1$. For other values of K and λ the new method is expected to be more powerful than the two existing methods.

The proposed HE method has similar power to the mean IBD test, when applied to affected sib pairs. However, to the extent that the mean IBD is only optimal under a multiplicative model of penetrances [Knapp et al., 1995], the currently proposed method is also not uniformly most powerful in all situations. It is possible to generalise the method to include dominance, by including the probability of complete IBD sharing as a second independent variable in the regression [Blackwelder and Elston, 1985]. This extended model is expected to be more powerful if the dominance variance of the QTL is substantial.

The proposed HE method uses linear regression even though the dependent variable is highly discrete, and this clearly violates the assumption of normality. Our simulation results indicate that, despite this violation of the normality assumption, the method is well behaved in terms of type 1 error rate. We have also assumed perfect IBD information which will not be the case with real data. The impact of imperfect IBD information will be to reduce the expected χ^2 statistics, and to increase the required sample size for ensuring the validity of standard asympotic theory for the test statistics.

The use of the proposed method requires that two parameters, namely K and λ be specified before the analysis. Risch [1990] has examined how these parameters are related to allele frequencies and penetrances, and how they influence to power of affected sib pair methods. Here, K and λ are required to obtain the optimal value of the weight α to be assigned to the product term XY in the dependent variable. Since the prevalence and sibling relative risk are basic genetic epidemiological parameters, it is hoped that their values will be known at least approximately prior to the linkage analysis. Minor mis-specification of these parameters is not expected to induce any spurious relationship between the dependent and independent variables in a linear regression analysis, so that the type 1 error rate should be unaffected. Furthermore, when K and λ are fairly large ($K > 0.1$, $\lambda > 2$), small misspecification in their values will cause little change in the weighting of the product term, so that there should be only a small decrease in the power of the test.

There are some limitations to our simulation studies. In particular, we have only examined large samples with a modest number of replicates. Therefore our statement about the robustness in type 1 error rate should be taken with caution, especially in small samples. When asymptotic theory is in doubt it would be desirable to obtain the empirical distribution of the test statistic by permutation procedures.

Our ability to construct an optimal regression method for binary traits speaks for the versatility and power of the approach advocated by Haseman and Elston over 30 years ago [Feingold, 2002]. The generalisation of this binary regression approach to general pedigree data should be possible, using the same theoretical framework as that which has been recently applied to quantitative traits [Sham et al., 2002].

Acknowledgements

Supported in part by grants from the Nutrition and Toxicology Research Institute Maastricht, the Netherlands (MPAZ), UK Medical Research Council grant G9901258 and US National Eye Institute grant EY-12562 (PCS), and United and The Leverhulme Trust, London, UK (JPR).

References

Abecasis GR, Cherny SS, Cookson WO, Cardon LR: 2002 Merlin-rapid analysis of dense genetic maps using sparse gene flow trees. Nat Genet 2002;30:97–101.

Allison DB, Fernandez JR, Heo M, Beasley TM: Testing the robustness of the new Haseman-Elston quantitative-trait loci-mapping procedure. Am J Hum Genet 2000;67:249–252.

Blackwelder WC, Elston RC: A comparison of sib-pair linkage tests for disease susceptibility loci. Genet Epidemiol 1985;2:85–97.

Cardno AG, Marshall EJ, Coid B, Macdonald AM, Ribchester TR, Davies NJ, Venturi P, Jones LA, Lewis SW, Sham PC, Gottesman II, Farmer AE, McGuffin P, Reveley AM, Murray RM: Heritability estimates for psychotic disorders: the Maudsley twin psychosis series. Arch Gen Psychiatry 1999;56:162–168.

Crittenden LB: An interpretation of familial aggregation based on multiple genetic and environmental factors. Ann N Y Acad Sci 1961;91:739–780.

Dawson DV, Kaplan EB, Elston RC: Extension to sib-pair linkage tests applicable to disorders characterized by delayed onset. Genet Epidemiol 1990;7:453–466.

Drigalenko E: How sib-pairs reveal linkage. Am J Hum Genet 1998;63:1242–1245.

Egan MF, Goldberg TE, Gscheidle T, Weirich M, Rawlings R, Hyde TM, Bigelow L, Weinberger DR: Relative risk for cognitive impairments in siblings of patients with schizophrenia. Biol Psychiatry 2001;50:98–107.

Elston RC, Buxbaum S, Jacobs KB, Olson JM: Haseman and Elston revisited. Genet Epidemiol 2000;19:1–17.

Holmans P: Asymptotic properties of affected sib-pair linkage analysis. Am J Hum Genet 1993;52:362–374.

Knapp M, Seuchter G, Baur MP: Linkage analysis in nuclear families I Optimality criteria for affected sib-pair tests. Hum Hered 1995;44:37–43.

Falconer DS: The inheritance of liability to certain diseases estimated from the incidence among relatives. Ann Hum Genet 1965;29:51–76.

Falconer DS, Mackay TFC: Introduction to quantitative genetics, ed 4. Essex, Longman, 1996.

Feingold E: Regression-based quantitative-trait-locus mapping in the 21st century. Am J Hum Genet 2002;1:217–222.

Fisher RA: The correlation between relatives on the supposition of Mendelian Inheritance. Trans R Soc Edinburgh 1918;52:399–433.

Forrest WF: Weighting improves the 'new Haseman-Elston' method. Hum Hered 2001;52:47–54.

Fulker DW, Cherny SS: An improved multipoint sib-pair analysis of quantitative traits. Behav Genet 1996;26:527–532.

Haseman JK, Elston RC: The investigation of linkage between a quantitative trait and a marker locus. Behav Genet 1972;2:3–19.

Kruglyak L, Daly M, Reeve-Daly M, Lander E: Parametric and nonparametric linkage analysis: A unified multipoint approach. Am J Hum Genet 1996;58:1347–1363.

Neale MC: QTL mapping with sib-pairs: the flexibility of Mx; in Spector TD, Snieder H, MacGregor AJ (eds): Advances in Twin and Sib-Pair Analysis. London, Greenwich Medical Media, 2000.

Neider A, Taneja S, Zeegers MP, Ostrer H: Genetic counseling for prostate cancer risk. Submitted, 2002.

Page WF, Braun MM, Partin AW, Caporaso N, Walsh P: Heredity and prostate cancer: a study of World War II veteran twins. Prostate 1997;33:240–245.

Risch N: Linkage strategies for genetically complex traits II. The power of affected relative pairs. Am J Hum Genet 1990;46:229–241.

Sham PC, Purcell S: Equivalence between Haseman-Elston and variance-components linkage analyses for sib-pairs. Am J Hum Genet 2001;68:1527–1532.

Sham PC, Purcell S, Cherny SS, Abecasis GR: Powerful regression-based quantitative-trait linkage analysis of general pedigrees. Am J Hum Genet 2002;71:238–253.

Snieder H: Genetic epidemiology of risk factors for coronary heart disease: A study of middle-aged twins. Enschede, CopyPrint, 2000.

StataCorp: Stata Statistical Software. Release 7.0. College Station, Stata Corporation, 2001.

Stuart A, Ord JK: Kendall's Advanced Theory of Statistics, vol 1, ed 6. London, Arnold, 1994.

Suarez BK, Rice JP, Reich T: The generalized sib-pair IBD distribution: its use in the detection of linkage. Ann Hum Genet 1978;42:87–94.

Visscher PM, Hopper JL: Power of regression and maximum likelihood methods to map QTL from sib-pair and DZ twin data. Ann Hum Genet 2001;65:583–601.

Wright FA: The phenotypic difference discards sib-pair QTL linkage information. Am J Hum Genet 1997;60:740–742.

Xu X, Weiss S, Wei LJ: A unified Haseman-Elston method for testing linkage with quantitative traits. Am J Hum Genet 2000;67:1025–1028.

Zeegers MP, Jellema A, Ostrer H: Empiric risk of prostate cancer for relatives of a prostate cancer patient. Cancer 2003;97:1894–1903.

Original Paper

Human Heredity

Hum Hered 2003;55:132–142
DOI: 10.1159/000072318

Information Perspectives of the Haseman-Elston Method

Fred A. Wright

Department of Biostatistics and Lineberger Cancer Center, University of North Carolina at Chapel Hill, Chapel Hill, N.C., USA

Abstract

The Haseman-Elston regression approach can be viewed as a special case in a larger framework in which independent sibling pairs are ascertained, perhaps based on their phenotypes. Using likelihood arguments in an idealized setting, expressions are obtained for the inherent linkage information contained in the phenotype distributions. The results provide insight and analytic results that apply to general phenotype distributions and can be used to judge the effects of various data transformations and selective sampling of extreme phenotypes.

Introduction

Haseman-Elston regression is in wide use as a means of mapping quantitative trait loci (QTLs) in sibling pairs, and the recent extensions to sibships [e.g. Wang and Huang, 2002] and more general pedigrees are likely to prove useful [Sham et al, 2002; see recent review by Feingold, 2002]. In the original Haseman-Elston approach [Haseman and Elston, 1972], the squared trait difference is regressed on the number of alleles shared identical-by-descent (IBD) at a marker locus. Wright [1997] showed that under bivariate normal assumptions the sum of the trait values provided additional information for detecting linkage – at the time I was unaware of the paper by Fulker and Cherny [1996] which contained many of the essentials of the argument. Several subsequent papers combining the squared differences and (mean corrected) squared sums have appeared and are reviewed in Feingold [2002].

Wright [1997] described the large-sample increase in expected LOD (ELOD) to be gained by using the trait sums in addition to the differences. A large-sample expression for the increase in ELOD was given, as well as a simple approximation to the proportionate increase under codominant models:

$$1 + \left(\frac{1-\rho_0}{1+\rho_0}\right)\left(\frac{1-\rho_2}{1+\rho_2}\right) \quad (1)$$

where ρ_j is the correlation between sibpair phenotypes when there are j alleles identical by descent at the QTL. As noted in Elston et al. [2000], a nearly identical formula appears in Gaines and Elston [1969], although space constraints precluded a full derivation in Wright [1997]. I use this occasion of the 30th anniversary of the original Haseman-Elston paper to describe the derivation in detail. In doing so, I introduce a view of the linkage problem in terms of the inherent information contained in the phenotype distributions, which differ as a function of IBD status at the QTL.

This view of the Haseman-Elston method is expansive and likelihood-based, focusing on regression as only one possible means of establishing linkage (although a useful and robust one). The approach offers concise analytic results and reveals some perhaps novel observations

KARGER

Fax + 41 61 306 12 34
E-Mail karger@karger.ch
www.karger.com

0001–5652/03/0553–0132$19.50/0

Accessible online at:
www.karger.com/hhe

Fred A. Wright
Department of Biostatistics and Lineberger Cancer Center
University of North Carolina at Chapel Hill
3107B McGavran-Greenberg Hall, CB #7420, Chapel Hill, NC 27599-7420 (USA)
Tel. +1 919 843 3655, Fax +1 919 966 3804, E-Mail fwright@bios.unc.edu

about the linkage approach. Although the treatment here is confined to sibling pairs, several of the lessons may be applicable to more general settings, and in particular the sections on mean-corrected phenotypes and selective sibpair genotyping suggest new approaches to identifying QTL linkage. Some of the ideas are anticipated in work by Elston et al. [2000], Forrest and Feingold [2000], Dudoit and Speed [2000], Liang et al., [2000], Wang et al. [2001] and others, but the approach described here may provide useful insights and a more unified perspective. Above all, this paper is an essay on the nature and sources of information in sibpair mapping. The figures and examples cover only a few relevant instances, and are not intended to be exhaustive.

Notation and a Likelihood Model

Let $\boldsymbol{f} = \{f_0, f_1, f_2\}$ represent the phenotype distributions when IBD = 0, 1, or 2 at the true QTL location, respectively. We will use the term 'density' to describe these distributions, although the treatment here applies equally well to discrete distributions of qualitative phenotypes. We use $y = \{y_1, y_2\}$ to denote the two phenotypes of the sibpair, and IBD(x) to represent the IBD status at genome position x. If the phenotype densities were known, the (natural) log-likelihood contribution for a single sibpair at a fully informative marker x would be

$$\begin{aligned} l(x) &= \log[P(y, IBD(x) = j)] \\ &= \log[P(IBD(x) = j)P(y \mid IBD(x) = j)] \\ &= \frac{1}{4}\log(f_0(y))\mathrm{Ind}_{[IBD(x)=0]} + \frac{1}{2}\log(f_1(y))\mathrm{Ind}_{[IBD(x)=1]} \\ &\quad + \frac{1}{4}\log(f_2(y))\mathrm{Ind}_{[IBD(x)=2]} \end{aligned} \quad (2)$$

where $\mathrm{Ind}_{[.]}$ is the indicator function. To facilitate comparison with later sections, we do not condition on the IBD status, but treat it as part of the data. A common approach used to gain intuition is to assume that the sample size is large enough that f_0, f_1, and f_2 can be considered known (see comments in Appendix 1). For most of this paper we will be concerned with the *precision information* (or simply 'information'), the slope of the expected log-likelihood (under the alternative hypothesis) at the QTL location for dense and informative markers. Although similar in spirit to the expected LOD score, the precision information often takes a simpler form and has a direct interpretation (Appendix 1) related to precision in estimating the QTL location once linkage has been established [Kong and Wright, 1994; Kruglyak and Lander, 1995]. Sen [1998] showed that with dense markers the median size (in Morgans) of $(1 - \alpha)100\%$ confidence intervals for a gene location is well-approximated by $2\log\alpha$/information, attributing the idea to A. Kong.

The Expected Log-Likelihood and Precision Information

We assume a true QTL location x^* and small recombination fraction θ between the QTL and informative marker x to the right of x^*. For a single sibpair we have

$$\begin{aligned} E_{x^*}(l(x)) &= \sum_{j=0}^{2} P(IBD(x^*) = j)E(l(x) \mid IBD(x^*) = j) \\ &\approx \frac{1}{4}E_{f_0}\{(1-\theta)^4\log(f_0(y)) + 4\theta\log(f_1(y))\} \\ &\quad + \frac{1}{2}E_{f_1}\{(1-\theta)^4 f_1(y)) + (1-\theta)^2 2\theta\log(f_0(y)) + (1-\theta)^2 2\theta\log(f_2(y))\} \\ &\quad + \frac{1}{4}E_{f_2}\{(1-\theta)^4\log(f_2(y)) + 4\theta\log(f_1(y))\} \end{aligned} \quad (3)$$

where the possible joint IBD states at x and x^* are considered from the four parental chromosomes. Throughout the paper '≈' signifies approximations accurate to terms of θ^2 and higher order (the exact expression involves, e.g., unlikely transitions from IBD = 0 to IBD = 2 over a short interval). Noting that $(1-\theta)^4 \approx 1 - 4\theta$, equation (3) can be rewritten

$$\begin{aligned} E(l(x)) \approx{}& \frac{1}{4}E_{f_0}\log f_0(y) + \frac{1}{2}E_{f_1}\log f_1(y) + \frac{1}{4}E_{f_2}\log f_2(y) \\ & - \theta[K(f_0, f_1) + K(f_1, f_0) + K(f_1, f_2) + K(f_2, f_1)] \end{aligned} \quad (4)$$

where

$$K(f, g) = E_f \log(f/g) = \int \log\left(\frac{f(y)}{g(y)}\right) f(y)dy$$

is the Kullback-Leibler (KL) 'distance' (although it is not in general symmetric) between f and g. For bivariate y, the integral is a double integral over the support of these densities. Defining $I(\boldsymbol{f}) = K(f_0, f_1) + K(f_1, f_0) + K(f_1, f_2) + K(f_2, f_1)$, equation (4) can be rewritten $E_{x^*}(l(x)) \approx E_{x^*}(l(x^*)) - \theta I(\boldsymbol{f})$, and because $\theta \approx x - x^*$, in the sense described above, the precision information is $\partial/\partial x\, E_{x^*}(l(x)) \approx I(\boldsymbol{f})$. Thus we have a neat correspondence between the summed Kullback-Leibler distances between the phenotype densities and the precision with which we can map the gene.

Additional Observations

Because the focus is on quantitative traits, I refer to f_0, f_1, and f_2 as densities. However, the derivations apply generally to any models for continuous or discrete phenotypes (the latter case involving sums rather than integrals in computing the KL distances). For many phenotype models $I(f)$ can be calculated analytically, without the need for simulation. Furthermore, the dependence on sample size is straightforward: for n independent sibling pairs, the information is $nI(f)$. Transformations of the phenotype data are easily accommodated by replacing the f's with the densities of the transformed observations, e.g. when the original bivariate sibpair observations are replaced with univariate observations such as the squared trait difference. We thus can use $I(f)$ as a clear criterion for assessing how much mapping information is retained after performing such simplifying transformations. From the treatment here it will be clear how to apply the results to other common types of relative pairs, which typically can have a maximum of 1 allele IBD and have different IBD transition rates than sibling pairs. Another aspect that cannot be explored here is that essentially all the results will apply to the very different biological setting of intercross experimental crosses, in which individuals (not pairs) share 0, 1, or 2 alleles IBD with a given parent strain.

The information is described below for several examples, repeatedly using the bivariate normal model for simplicity and because it often results in closed-form analytic expressions. However, any phenotype densities/distributions may be used.

Information in the Differences vs. the Sums

For the sibpair bivariate normal model of Wright [1997] we can consider the dependence of the information on the correlation coefficients. We assume $E(y_1) = E(y_2) = \mu$, $\text{var}(y_1) = \text{var}(y_2) = \xi^2$, regardless of IBD status. For $\text{IBD}(x^*) = j$, the differences $D = y_1 - y_2$ have mean 0, variance $2\xi^2(1 - \rho_j)$. Redefining the 'phenotype' as D, for $\text{IBD}(x^*) = j$ we have $f_j = N(0, 2\xi^2(1 - \rho_j))$, and the information for D and S are (Appendix 2 shows the calculation of KL distances between normal densities)

$$I_D = -2 + \frac{1}{2}\left\{\frac{1-\rho_0}{1-\rho_1} + \frac{1-\rho_1}{1-\rho_0} + \frac{1-\rho_1}{1-\rho_2} + \frac{1-\rho_2}{1-\rho_1}\right\},$$

$$I_S = -2 + \frac{1}{2}\left\{\frac{1+\rho_0}{1+\rho_1} + \frac{1+\rho_1}{1+\rho_0} + \frac{1+\rho_1}{1+\rho_2} + \frac{1+\rho_2}{1+\rho_1}\right\}. \qquad (5)$$

The information increases rapidly as the correlation coefficients increase. Using $\rho = (\rho_0, \rho_1, \rho_2)$ to denote the vector of correlation coefficients, we have $I_D = 0.032$ when $\rho = (0.0, 0.15, 0.3)$ and $I_D = 0.225$ when $\rho = (0.0, 0.3, 0.6)$. In this instance, when using the trait differences, doubling the correlations results in a 7-fold increase in mapping precision.

By the independence of D and S [Pitman, 1939], the total information using the complete bivariate data is $I_{D,S} = I_D + I_S$ (Appendix 2), and so $I_{D,S}/I_D = 1 + I_S/I_D$. When ρ_1 is intermediate to ρ_0 and ρ_2,

$$I_D = \frac{(\rho_1 - \rho_0)^2}{(1-\rho_2)(1-\rho_0)} \text{ and } I_S = \frac{(\rho_1 - \rho_0)^2}{(1+\rho_2)(1+\rho_0)}.$$

This simplification clarifies that the information depends not only on the differences between the correlations, but also on the absolute correlations (through the denominator). Thus while $I_D = 0.225$ when $\rho = (0.0, 0.3, 0.6)$, $I_D = 0.333$ (nearly 50% greater) when each of the correlation coefficients is increased by 0.1, or $\rho = (0.1, 0.4, 0.7)$. The simplified form for I_D and steps in the calculation of I_S/I_D are provided in Appendix 2, finally giving equation (1).

It is useful to note that when $\rho_1 = (\rho_0 + \rho_2)/2$, the slope of the Haseman-Elston regression line is $\xi^2(\rho_0 - \rho_2)$, while in the original derivation [Haseman and Elston, 1972] is shown to be the negative of the genetic variance. Because the total phenotypic variance is ξ^2, it follows that the heritability (genetic variance/total variance) is $H^2 = (\rho_2 - \rho_0)$. Comparing this result to the simplified expression for I_D above, we obtain

$$I_D = \frac{1}{4} H^2 \frac{(\rho_2 - \rho_0)}{(1-\rho_2)(1-\rho_0)}.$$

For small ρ_2 (as with a weak effect of a complex trait QTL),

$$I_D \approx \frac{1}{4}\{H^2\}^2.$$

A similar relationship is easily derived for I_S. The heritability is a more familiar quantity than the information measures described here, so it is reassuring to see such a simple relationship.

Transformations and Regression

For both D and S, it is the differing variances by IBD status that ultimately provide the linkage information, although in common practice the squared differences $D^2 = (y_1 - y_2)^2$ and mean-corrected sums $(S - 2\mu)^2$ are often used instead. It is the *expectations* of these latter quantities that differ by IBD status. We will show below that these two transformations are fully efficient, in the sense that the precision information for D is the same as that for D^2, and the information for S is the same as that for $(S - 2\mu)^2$. However, the question remains whether a simple univariate transformation of the original *bivariate* data can retain a large portion of the total linkage information. Proposed transformations include weighted averages of D^2 and $(S - 2\mu)^2$ [Xu et al., 2000; Forrest 2001] and the 'Haseman-Elston Revisited' (*HER*) transformation $(y_1 - \mu)(y_2 - \mu)$ [Elston et al., 2000; also see Drigalenko, 1998] (see technical comments in Appendix 3). Here μ is known because the f densities are considered known.

Efficiency of the HE Revisited Approach

Figure 1 compares the information per sibpair for various transformations for our example with $\rho = (0, 0.3, 0.6)$, and with population trait mean $\mu = 5.0$. The inherent precision information based on the likelihood is presented for transformations D, S and the Haseman-Elston revisited $HER(\mu') = (y_1 - \mu')(y_2 - \mu')$ over a range of values μ'. The figure illustrates that the *HER* transformation contains maximum information (calculated numerically) for $\mu' = \mu$, consistent with results from Wang et al. [2001] (hereafter '*HER*' without the offset argument will signify $HER(\mu)$). Moreover, the information for *HER* exceeds I_D. However, the information is always less than the full information $I_{D,S}$.

The upper panels of figure 2 show contour plots of the f_0, f_1 and f_2 for the bivariate normal model with $\mu = 0$, $\xi^2 = 1$, and the same correlations $\rho = (0, 0.3, 0.6)$. The lower panels show f_0, f_1 and f_2 after transformations D, S and *HER*, respectively. It is apparent that even for a QTL with strong effects the densities do not differ greatly by IBD status, and that the D densities are more disparate than the S densities. It may seem that the *HER* density (lower right panel) shows less disparity by IBD status than either D or S. However, because D and S differ only in variance by IBD status, when computing the KL distance between, say, f_0 and f_1, for the transformed phenotype $y = D$, the integrand $\log(f_0(y)/f_1(y))f_0(y) < 0$ for y near zero is partly cancelled by the positive values for the same quantity in the tails. The product-normal *HER* transformation, in contrast, shows a steady shift to the right with increasing IBD status.

General Transformations

In the discussion above, we described the information based on D and S directly, although D^2 and the mean-corrected squared sum $(S - 2\mu)^2$ are often used instead as the response variables in Haseman-Elston regression. A more general question naturally arises: can transformations of a phenotype possibly *increase* the precision information? The answer is no, for reasons described in some detail in Appendix 3 (although transformations might be

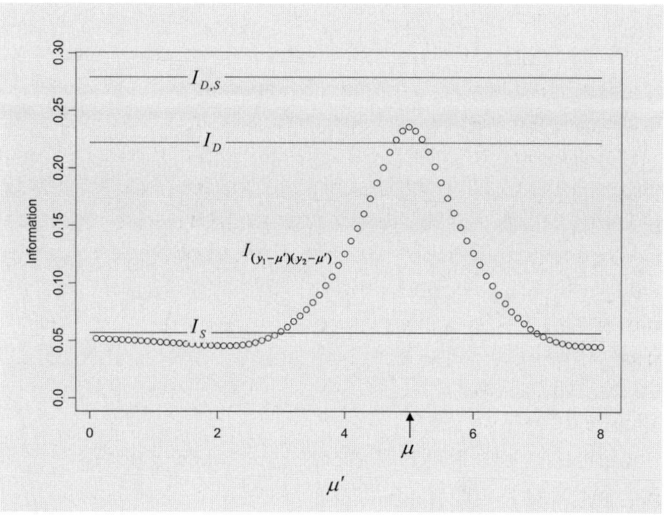

Fig. 1. Individual and combined precision information for D and S, as well as $HER(\mu') = (y_1 - \mu')(y_2 - \mu')$ for a variety of offset values μ'. The information for $HER(\mu')$ is maximized at the population mean μ.

advantageous in terms of power or robustness when using a non-likelihood procedure). This result is intuitively satisfying, for otherwise we would somehow be gaining 'free' information not contained in the likelihood. As described in the Appendix, one-to-one transformations result in no loss of information, while non-monotone transformations will in general result in loss of information. There is an exception, however. For transformations $r(y)$ in which $r(y^A) = r(y^B)$ implies that $f_j(y^A) = f_j(y^B)$ (here y^A and y^B are two different phenotype values), then there is no such loss of information. For example, in our normal model, $D \sim N(0, 2\xi^2(1 - \rho_j))$ when $\text{IBD}(x^*) = j$. Applying the transformation D^2 retains all of the information from the original D, because the density of any realized value $D = -d$ is the same as that of $D = +d$, and both have the same transformed value d^2. A similar argument applies to $(S - 2\mu)^2$, in which there is no loss compared to S (although there *is* information loss if the sum is corrected by any value other than 2μ). Thus we have a direct information-based justification for these transformations, or at least that they are not harmful.

Regression

We now consider the technique of regression as a means to detect linkage to the QTL with D^2 as the response variable. It is reasonable to expect that some information is lost when performing simple least-squares regression, because typical assumptions of normality and

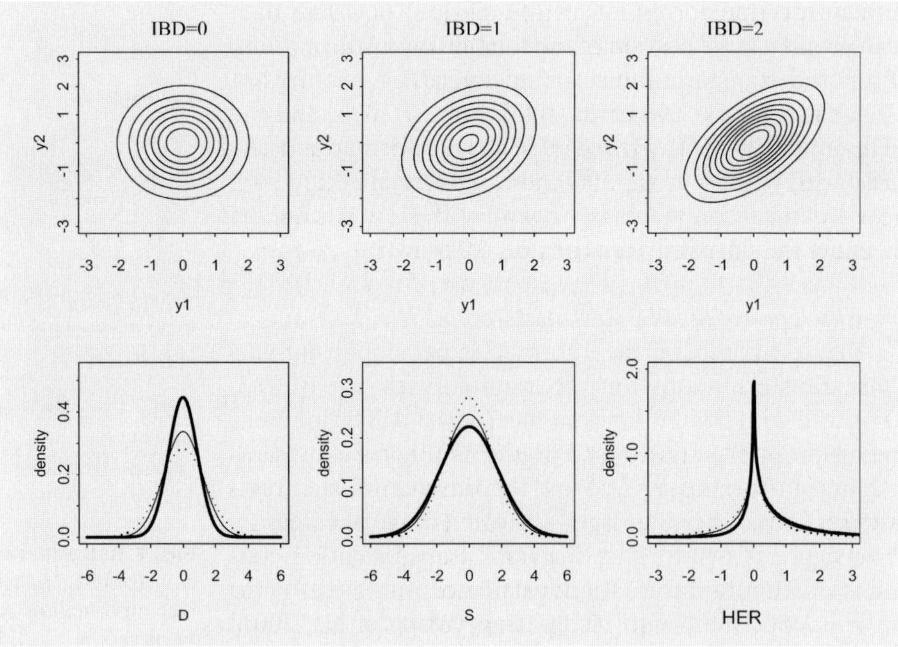

Fig. 2. Upper panels: contours of the bivariate densities f_0, f_1, and f_2, for the example with $\rho = (0.0, 0.3, 0.6)$. Lower panels: the corresponding densities of D, S, and HER, for IBD(x^*) taking values 0 (dotted), 1 (thin lines), and 2 (bold lines).

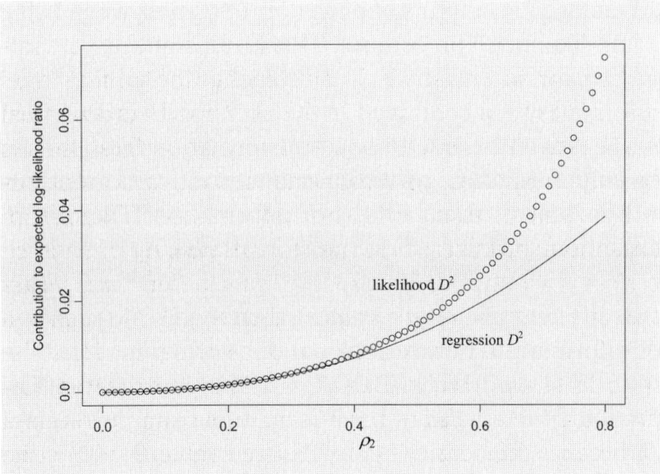

Fig. 3. Expected contribution of a sibpair to the expected natural log-likelihood ratio at the QTL, for the bivariate normal model and assuming $\rho_0 = 0$, $\rho_1 = \rho_2/2$. For modest ρ_2 values, the expectation from the full likelihood (open circles) exceeds an analogous regression-based quantity (line) by only a small amount.

variance homogeneity do not hold. However, it is difficult to express the regression in terms of precision information per sibling pair. Thus, for this subsection we consider the distribution of regression coefficients at the QTL vs. an unlinked location. From these distributions we obtain an expected log-likelihood ratio (i.e., an ELOD using natural logs). Thus these expected log-likelihoods can be used to compare the simple linear regression approach to the full likelihood approach. The calculations are given in Appendix 4. Figure 3 plots the contributions for a wide range of ρ_2 values, where $\rho_0 = 0$ and $\rho_1 = \rho_2/2$. The expected log-likelihood ratio contributions per sibling pair are remarkably similar for the likelihood vs. regression techniques for ρ_2 up to about 0.4, above which the difference becomes substantial. The fact that unweighted simple regression could perform so well may be a bit surprising, considering that the response variable is so highly non-normal (proportional to χ_1^2 for each IBD value). However, the regression slope depends essentially on the sample means for each IBD value, and it can be shown for these data that the sample mean is the maximum likelihood estimator of the population mean (Appendix 4). When ρ_1 is not intermediate to ρ_0 and ρ_2, then the regression approach will further lose precision compared to that of the likelihood.

Mean-Corrected Phenotypes: Family Mean vs. Global Mean

Wang et al. [2001] described transformations of the paired phenotypes of the form $HER(\mu') = (y_1 - \mu')(y_2 - \mu')$, and noted that the D^2 and the HER transformations are equivalent or proportional to two special cases of this transformation, for $\mu' = (y_1 + y_2)/2$ and $\mu' = \mu$, respectively. These two choices correspond to the family mean (for a two-sib family) and the global mean of the phenotypes, and it has been shown that using the family mean $\mu' = (y_1 + y_2)/2$ in $HER(\mu')$ (equivalent to using D^2) is often advantageous, especially for multiple-sib families when there are family-specific environmental or genetic effects [Palmer et al., 2000; Wang et al. 2001]. This is understandable if we consider simple phenotype models involving random family-specific effects,

$$y_1 = \mu + \eta_k + (QTL\ effect)_1 + \varepsilon_1$$
$$y_2 = \mu + \eta_k + (QTL\ effect)_2 + \varepsilon_2,$$

where η_k is the family-specific effect for sibship k. With the QTL effects and random error terms standardized to have mean zero, the family mean $(y_1 + y_2)/2$ is an unbiased estimate of $\mu + \eta_k$, and so subtracting this amount from each phenotype may better reveal variability due to the QTL effect alone. However, the fact that the family mean is unbiased must be weighed against the variability in estimation, so that using the fixed value μ instead may be superior if var(η) is small. These competing considerations are quite familiar from the mixed-model and random-effects model literature, in which estimates of a quantity are 'shrunk' toward a global mean, and may be preferred in terms of expected mean-squared error [McCulloch and Searle, 2001, p 50]. In our setting, such a shrinkage estimate would be the weighted average

$$\mu' = (1 - \delta)\left(\frac{y_1 + y_2}{2}\right) + \delta\mu$$

where δ is near 1 if var(η) is small, and near zero if var(η) is large. We must defer the practical details of estimating the random-effect variance to future work. However, the idea is straightforward and compelling, perhaps offering improvements over current approaches. For realistic situations δ should depend on the size of the sibship, to reflect the relative precision of estimating the family-specific effect within each family. From a prediction standpoint the theory of mixed-models would prescribe a best linear unbiased estimator of δ [McCulloch and Searle, 2001, p 50], but this estimator is apparently not optimal in terms of mapping precision. A counter-example is our bivariate normal example with $\rho = (0, 0.30, 0.60)$, shown in figure 4 for δ ranging from 0 (the special case equivalent in information to D^2) to 1 (equivalent to HER). Note that for these two extreme δ's the y-axis is equivalent to I_D and $I_{(y_1-\mu)(y_2-\mu)}$, respectively, from figure 1. In this situation there is truly no family-specific effect, so clearly the best estimate for η_k is zero for each sibpair, suggesting the use

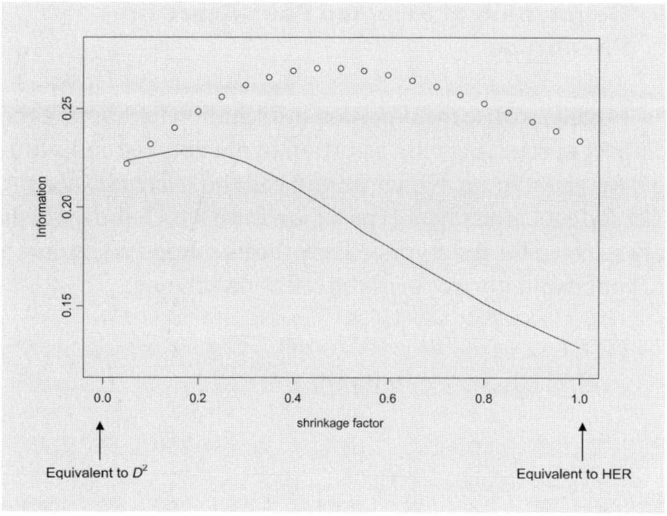

Fig. 4. Using a shrinkage estimate in $HER(\mu')$ (i.e., a nonconstant μ' that is a weighted average of the sibpair mean and the global mean μ) can improve the information, even in the absence of family-specific effects (open circles). When family-specific effects are added (line), the optimal shrinkage factor is small, so that the best μ' is near the sibpair mean.

of $\delta = 1$. However, the information (fig. 4) is in fact maximized near $\delta = 0.45$, rising to within 3% of the theoretical maximum information $I_{D,\ S}$ from figure 1. Thus in this case the use of shrinkage has recovered much of the information lost by using HER instead of the full likelihood, and it is not clear how widespread this phenomenon is, or how applicable to the phenotype models. The data generated for this example was then modified by adding a $N(0,1)$ family-specific effect to each sibpair (recall that the original global variance was $\xi^2 = 1.0$). As expected, figure 4 shows that the information is now maximized at a smaller value, near $\delta = 0.2$. The convergence of the two curves at $\delta = 0$ occurs because the η_k term cancels when using $\mu' = (y_1 + y_2)/2$ in $HER(\mu')$.

While the special cases of full shrinkage and no shrinkage have been examined [Wang et al, 2001] the general use of shrinkage has not been examined carefully [although see fig. 2 in Palmer et al., 2000]. In practical mapping settings, one possibility is to treat δ as an unknown parameter, choosing it to maximize, e.g., the LOD score at each genomic position. Identifying an optimal shrinkage proportion remains an important unsolved problem, as it may effectively account for important components of phenotypic variance while retaining the simple and robust character of Haseman-Elston regression.

Genotyping of Selected Pairs Based on Phenotype

Finally, we use the precision information to examine the impact of selecting only a portion of phenotyped individuals for genotyping, which we will call the *selection scheme.* Let R denote the phenotype region from which individuals are selected for genotyping. Here the likelihood is similar to (2) but explicitly involves the selection scheme:

$$P(data; x) = P(y, IBD(x) = j \mid y \in R)$$
$$= P(IBD(x) = j \mid y \in R)\, P(y \mid IBD(x) = J, y \in R).$$

Define

$$z_j = P(IBD(x) = j \mid y \in R),$$

and

$$h_j(y) = P(y \mid IBD(x^*) = j, y \in R) = \frac{f_j(y)}{\int_R f_j(y)dy}$$

for the true unknown location x^*. Then the log-likelihood for a sibpair with phenotype y and IBD$(x) = j$ is

$$l(x) = \log(P(y, \mathrm{IBD}(x) = j \mid y \in R)) = \log(z_j\, h_j(y)) \quad (6)$$

and

$$l(x^*) - l(x) \approx$$

$$\left[\log\left(\frac{h_0(y)}{h_1(y)}\right) + \log\left(\frac{z_0}{z_1}\right)\right] \mathrm{Ind}_{[IBD(x^*) = 0,\, IBD(x) = 1]} +$$

$$\left[\log\left(\frac{h_1(y)}{h_0(y)}\right) + \log\left(\frac{z_1}{z_0}\right)\right] \mathrm{Ind}_{[IBD(x^*) = 1,\, IBD(x) = 0]} +$$

$$\left[\log\left(\frac{h_1(y)}{h_2(y)}\right) + \log\left(\frac{z_1}{z_2}\right)\right] \mathrm{Ind}_{[IBD(x^*) = 1,\, IBD(x) = 2]} +$$

$$\left[\log\left(\frac{h_2(y)}{h_1(y)}\right) + \log\left(\frac{z_2}{z_1}\right)\right] \mathrm{Ind}_{[IBD(x^*) = 2,\, IBD(x) = 1]} \quad (7)$$

In order to take expectations when the QTL location is x^*, we note that

$$P(IBD(x^*) = 0, IBD(x) = 1 \mid y \in R)$$
$$= P(IBD(x^*) = 0 \mid y \in R) P(IBD(x) = 1 \mid IBD(x^*) = 0)$$
$$\approx z_0 4\theta$$

and the probabilities for the other transition IBD terms in equation (6) are computed similarly. The rescaled h_j are proper densities, integrating to 1 over the region R and defined as zero elsewhere. We can now compute $E_{x^*}(l(x^*) - l(x) \mid y \in R) \approx \theta I_R(f)$, where $I_R(f)$ is the information under the selection scheme,

$$I_R(f) = \{4z_0K(h_0, h_1) + 2z_1[K(h_1, h_2) + K(h_2, h_1)] + 4z_2K(h_2, h_1)\}$$
$$+ \{4z_0\log(z_0/z_1) + 2z_1[\log(z_1/z_0) + \log(z_1/z_2)] + 4z_2 \log(z_2/z_1)\} \quad (8)$$

where the terms inside the braces are denoted $I_R(\boldsymbol{h})$ and $I_R(\boldsymbol{z})$, respectively. Thus the information divides cleanly into a portion consisting of a weighted average of the distances among the h densities, and a portion involving the departure of the IBD sharing probabilities $\{z_0, z_1, z_2\}$ from their 'null' values of {1/4, 1/2, 1/4} under no selection (or equivalently, their values at an unlinked locus, regardless of selection). It is easy to verify that when no selection occurs (i.e. R is the entire range of y), equation (6) is equivalent to the unselected log-likelihood (2), and from (8) we see that in this case $I_R(\boldsymbol{f}) = I_R(\boldsymbol{h}) = I(\boldsymbol{f})$, as each $h_j = f_j$ and terms in $I_R(\boldsymbol{z})$ cancel each other.

The log-likelihood in (6) is also equally applicable to qualitative traits, and in particular affected sibpair mapping. Recall that $y = \{y_1, y_2\}$ is the pair of sib phenotypes, so that if affected status is coded as 0 (unaffected) or 1 (affected), all sampled pairs have $y = \{1,1\}$. Here the rescaled h distributions will all have mass 1 at $y = \{1,1\}$, and the distances among them will be zero, while the z_j's correspond to the familiar notation for IBD sharing probabilities among affected sibpairs. The $\boldsymbol{f}$ distributions are the probabilities for the three possible joint phenotypes {0,0}, {0,1}, {1,1} (assuming that {0,1} is equivalent to {1, 0} for each IBD value at the disease locus).

Discussion

We have explored how differences among phenotype densities/distributions, conditioned on IBD status, determine the mapping precision information. The results are expressed for general distributions, and can be used to gain insight into the efficiency of simplifying transformations, and serve as a bridge between the traditional Haseman-Elston setting and methods developed for qualitative traits and affected-sibling pairs. While the information perspective can be used to judge the benefits of full use of the data vs. transformed data or other simplifications, similar efforts are needed to understand the cost of estimating numerous parameters under full specification, or of possible model misspecification. The use of regression (as opposed to the likelihood described here) may also provide additional robustness to violations of the phenotype model. Here likelihood is primarily used for motivation and intuition, suggesting possible improvements that may later be implemented using more robust procedures.

One natural extension to the information approach is to consider models for sibships of arbitrary size, with densities for all possible joint IBD states and with dimension equal to the number of sibs in the family. Although fitting such a large phenotype model may be impractical, the exercise might be useful in understanding the impact of such dimensionality reduction techniques implicit in sibship score tests [Wang and Huang, 2002] or the effect of treating sibling pairs as independent for testing purposes [Elston et al., 2001]. It may also point the way to yet more powerful robust statistics for detecting QTL linkage in pedigrees.

To be useful in data analysis, the selected log-likelihood (6) must be extended to include estimation of the $\boldsymbol{f}$ distributions and account for uncertainties in IBD status based on marker genotypes. The likelihood is simple in form and serves as a clear means to unify the sibpair mapping of quantitative and qualitative traits, so it is perhaps surprising that it is not commonly used. Several recent reports have described unifications of the two techniques, [Liang et al., 2000; Dudoit and Speed, 2000; Forrest and Feingold, 2000; Chang et al., 2002]. The work by Dudoit and Speed [2000] appears closest to the current treatment, explicitly invoking a conditional likelihood in a score test for linkage.

We can use equation (6) to directly examine the inherent information provided by each *individual* phenotype value for continuous traits. To do so, we take the limit of $I_R(\boldsymbol{f})$ as the sampling region R is reduced to a single point y. Because of the rescaling, $I_R(\boldsymbol{h}) = 0$. Denoting

$$\bar{f}(y) = \frac{1}{4}f_0(y) + \frac{1}{2}f_1(y) + \frac{1}{4}f_2(y),$$

we obtain

$$I_y(\boldsymbol{f}) = \log\left(\frac{f_0(y)}{2f_1(y)}\right)\left(\frac{f_0(y)-f_1(y)}{\bar{f}(y)}\right) + \log\left(\frac{f_2(y)}{2f_1(y)}\right)\left(\frac{f_2(y)-f_1(y)}{\bar{f}(y)}\right) \quad (9)$$

(see comment in Appendix 5). From this simple formula we can examine $I_R(\boldsymbol{f})$ for the bivariate normal example described in figure 2, with $\rho = (0, 0.3, 0.6)$. Figure 5A plots these information contours, from which we can immediately see that the information is lowest in the center, and increases toward each corner of the figure. In other words, the information is high for extreme discordant pairs (pairs with very different phenotypes on the upper left and lower right), as well as those whose phenotypes are highly concordant but are extreme (upper right and lower left). The value of genotyping such extreme pairs has been studied by others in a less formal manner [Risch and Zhang, 1995; Zhang and Risch, 1996; Dudoit and Speed, 2000; Liang et al., 2000].

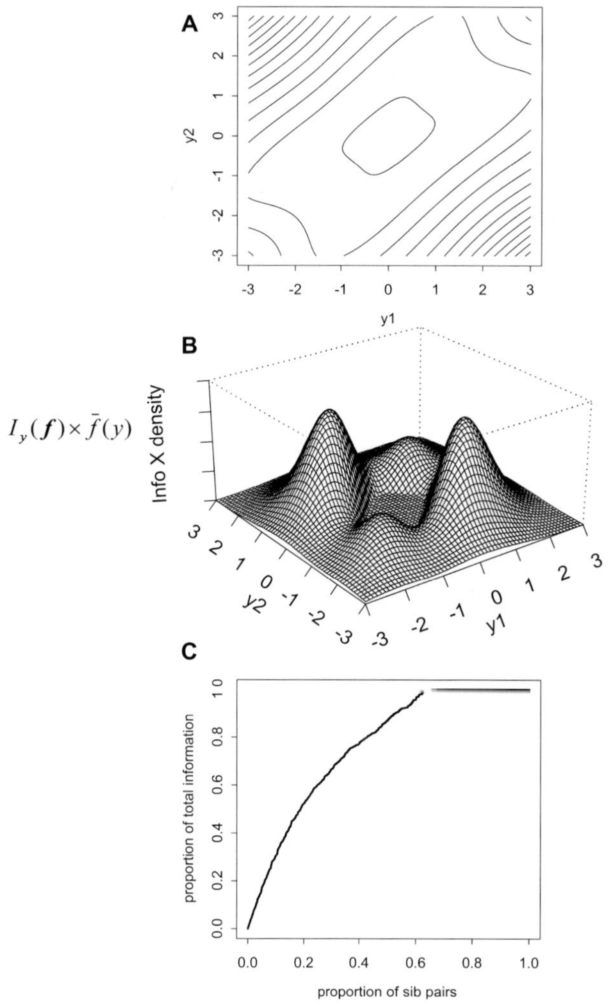

Fig. 5. Consequences of genotyping extreme sibpairs. **A** Contours of $I_y(\boldsymbol{f})$ for the bivariate normal example with $\rho = (0, 0.3, 0.6)$. The information increases toward the corners of the plot, but more quickly for discordant pairs (upper left – lower right). **B** $I_y(\boldsymbol{f}) \times \bar{f}(y)$ indicates the average informativeness of y in unselected samples, which again shows that most of the information arises from discordant and concordant pairs, but that the information for very extreme pairs is limited by the availability of such pairs. **C** The proportion of total information for most informative portions of sibpairs.

In addition, it is straightforward to compute how each phenotype pair contributes to the overall information when no selection is performed. Although the information contributed by a single pair with phenotype y is $I_y(\boldsymbol{f})$, the relative information contributed by all such pairs is $I_y(\boldsymbol{f}) \times \bar{f}(y)$, because such pairs occur with frequency $\bar{f}(y)$. A perspective plot of $I_y(\boldsymbol{f}) \times \bar{f}(y)$ vs. y is provided for the same example in figure 5B. Again, the discordant pairs and concordant pairs provide most of the total information (the four peaks), but now we see that the most extreme pairs (the very extreme corners of the plot) do not provide much information on average in unselected samples, because they occur so seldom. Also, the discordant pairs have higher peaks than the concordant pairs, providing several times as much average information. Finally, we can easily examine how the total information in large, unselected samples is divided among the sibpairs. Figure 5C shows the cumulative information when only a most informative proportion of the sibpairs are used, with the proportion varying from 0 to 1.0. For this example, the cumulative information provided by the 30% most informative sibpairs accounts for ~65% of the precision information. Note that this observation addresses a different issue than the truly dramatic gains in mapping efficiency achievable in a huge phenotyped population, from which the most extreme can be genotyped (e.g. fig. 5A and Risch and Zhang [1995]).

Acknowledgments

Supported in part by NIH GM58943. Jianhua Hu provided assistance with Appendix 3.

Appendix 1: Asymptotics

Assuming Phenotype Densities Known

Although the formal arguments for this assumption can be quite involved, requiring the use of weak convergence, essentially this results from the fact that with increasing sample size n, any parameters to be estimated in specifying the phenotype densities are accurate to order $1/\sqrt{n}$. According to the arguments of Kong and Wright [1994], and similar results for affected sibpairs in Kruglyak and Lander (1995), the number of crossovers in a plausible interval likely to contain the gene is $O_p(1)$, involving a proportion of sibpairs of order $1/n$. The net effect is that any reasonable approach to estimating the model phenotype parameters (such as profile likelihood) will result in a log-likelihood in the vicinity of the maximum that differs little from the 'true' likelihood had the densities been known. Several results in this paper technically require smoothness and other conditions on the phenotype densities. The expected log-likelihood under the null hypothesis does not appear in the information definition because it is nearly constant in the vicinity of the QTL in most settings.

Median Size of Confidence Intervals

Sen [1998] explored in depth a heuristic he attributes to Augustine Kong, in which the exponentiated expected log-likelihood in the vicinity of the gene (essentially double exponential) is treated as a posterior for the purpose of finding a likely size of a confidence interval. By the properties of the double-exponential, the width is $2\log\alpha/$ slope for a $(1-\alpha)100\%$ confidence interval, where 'slope' is our precision information, defined as a limiting slope as one approaches the QTL location from below (to account for a discontinuity at the QTL location). This heuristic was compared to Bayesian intervals and the Kruglyak and Lander procedure [1995], using profile likelihood. The result was that the procedure gives a remarkably accurate estimate of the *median* size of a confidence interval when mapping a gene using dense and informative markers.

Appendix 2: Kullback-Leibler Information for D, S

KL Distance between Normal Densities

We have $f_0 = N(\mu_0, \sigma_0^2), f_1 = N(\mu_1, \sigma_1^2)$ and

$$K(f_0, f_1) = E_{f_0} \log\left(\frac{f_0(Y)}{f_1(Y)}\right)$$

$$= E_{f_0}\left(\frac{1}{2}\log\frac{\sigma_1^2}{\sigma_0^2} - \frac{1}{2}\left(\frac{(Y-\mu_0)^2}{\sigma_0^2} - \frac{(Y-\mu_1)^2}{\sigma_1^2}\right)\right)$$

$$= \frac{1}{2}\log\frac{\sigma_1^2}{\sigma_0^2} - \frac{1}{2}\left(1 - \frac{\sigma_0^2 + (\mu_0-\mu_1)^2}{\sigma_1^2}\right).$$

Applying this result to I_D and I_S is straightforward, substituting the appropriate means and variances, and noting that the log terms cancel in the overall information.

Kullback-Leibler Information Adds for Independent D, S

We use f_{jD} and f_{jS} to denote the densities of D and S for IBD(x^*) = j, and the density of the original bivariate observation is $f_{jD}\,f_{jS}$. We have

$$K(f_0, f_1) = K(f_{0D}f_{0S}, f_{1D}f_{1S}) = E_{f_{0D}f_{0S}} \log\left(\frac{f_{0D}f_{0S}}{f_{1D}f_{1S}}\right)$$

$$= E_{f_{0D}f_{0S}} \log\left(\frac{f_{0D}}{f_{1D}}\right) + E_{f_{0D}f_{0S}} \log\left(\frac{f_{0S}}{f_{1S}}\right).$$

(to avoid confusion of the variable with the subscripts we have omitted the variable, e.g., $f_{0D}(D)$). The first term is

$$E_{f_{0D}f_{0S}} \log\left(\frac{f_{0D}}{f_{1D}}\right) = \int_S \int_D \log\left(\frac{f_{0D}}{f_{1D}}\right) f_{0D}f_{0S}\, dD\, dS$$

$$= \int_S K(f_{0D}, f_{1D})\, f_{0S}\, dS = K(f_{0D}, f_{1D}),$$

and similarly for the second term. Applying this result to all the terms in $I(f)$ gives the desired final result.

Deriving Equation (1)

If we let $\alpha = 1 - \rho_1$, assume $\rho_1 = (\rho_0 + \rho_2)/2$, and let $\beta = \rho_1 - \rho_0$, then we can rewrite

$$I_D = -2 + \frac{1}{2}\left\{\frac{\alpha+\beta}{\alpha} + \frac{\alpha}{\alpha+\beta} + \frac{\alpha}{\alpha-\beta} + \frac{\alpha-\beta}{\alpha}\right\},$$

which can eventually be reduced to

$$I_D = \frac{\beta^2}{(\alpha-\beta)(\alpha+\beta)}$$

or

$$I_D = \frac{(\rho_1 - \rho_0)^2}{(1-\rho_2)(1-\rho_0)}.$$

A similar expression can be obtained for I_S using $\gamma = 1 + \rho_1$, giving finally

$$\frac{I_S}{I_D} = \frac{(\alpha-\beta)(\alpha+\beta)}{(\gamma-\beta)(\gamma+\beta)} = \frac{(1-\rho_2)(1-\rho_0)}{(1+\rho_0)(1+\rho_2)},$$

and

$$\frac{I_{D,S}}{I_D} = 1 + \frac{I_S}{I_D}$$

implies the result.

Appendix 3: Transformations

It is natural to use the $I(f)$ criterion to consider whether univariate transformations of the phenotypes can retain all of the information. In a strict sense the answer is yes, because the densities f_0, f_1, and f_2 of the original bivariate data are themselves such mappings, which is simply another way of stating that the full likelihood approach is efficient. However, for independent sibpairs and under the bivariate normal assumptions, none of the commonly used transformations (*D, S, HER*) retain 100% of the linkage information. This phenomenon is closely related to sufficient statistics for the correlation coefficients ρ_j, the only parameters that differ by IBD status. Considering the bivariate normal likelihood for n_j independent sibpairs, it can be easily seen via the factorization theorem [Cox and Hinkley, 1974, p 21] that the minimal sufficient statistic for ρ_j is the two-dimensional quantity

$$\sum_i^{n_j} (y_{1_i} - \mu)(y_{2_i} - \mu),\ \sum_i^{n_j} y_{1i}^2 + \sum_i^{n_j} y_{2i}^2$$

where each sibpair is indexed by i. None of the common transformations are equivalent to this statistic, although the sum or mean of the HER transformation $(y_1 - \mu)(y_2 - \mu)$ (the first part of the sufficient statistic) has appealing properties related to conditional sufficiency for 'nearly' ancillary statistics [Cox and Hinkley, 1974, example 2.30].

General Results on Loss of Information

One-to-one phenotype transformations $w = r(y)$ result in no change in information. This should be apparent for discrete phenotype distributions, where each $f_j(y)$ is replaced in the KL distances by an identical $g_j(w) = f_j(r^{-1}(w))$. For continuous phenotype densities, define the inverse transformation $y = s(w)$. A standard calculus result is

$$g(w) = f[r^{-1}(w)]\frac{dr^{-1}(w)}{dw} = f[s(w1X)]\frac{d\,s(w)}{dw}.$$

Thus

$$K(g_0, g_1) = E_{g_0}\log\left(\frac{g_0(W)}{g_1(W)}\right) = E_{g_0}\log\left(\frac{f_0[s(W)]\frac{ds(W)}{dW}}{f_1[s(W)]\frac{ds(W)}{dW}}\right)$$

$$= \int \log\left(\frac{f_0[s(w)]}{f_1[s(w)]}\right) f_0[s(w)]\frac{ds(w)}{dw}\,dw,$$

and by another change of variables this is

$$\int \log\left(\frac{f_0(y)}{f_1(y)}\right) f_0(y)\,dy = K(f_0, f_1).$$

For transformations that are not one-to-one, a sketch of the argument follows. For two different phenotype values y_A and y_B, the contribution to $K(f_0, f_1)$ is

$$K_1 = \log\left(\frac{f_0(y^A)}{f_1(y^A)}\right) f_0(y^A) + \log\left(\frac{f_0(y^B)}{f_1(y^B)}\right) f_0(y^B).$$

Suppose we have a transformation/mapping $w = r(y)$ such that $r(y^A) = r(y^B)$, but is one-to-one for all other y. The contribution to the new KL distance between $g_0(w)$ and $g_1(w)$ is

$$K_2 = \log\left(\frac{f_0(y^A) + f_0(y^B)}{f_1(y^A) + f_1(y^B)}\right)(f_0(y^A) + f_0(y^B))$$

and we need only show that $K_1 \geq K_2$. To do so, we construct a random variable V which assumes the value $f_0(y^A)/f_1(y^A)$ with probability

$$\frac{f_1(y^A)}{f_1(y^A) + f_1(y^B)},$$

and the value $f_0(y^B)/f_1(y^B)$ with probability

$$\frac{f_1(y^B)}{f_1(y^A) + f_1(y^B)}.$$

Now

$$E(V) = \frac{f_0(y^A) + f_0(y^B)}{f_1(y^A) + f_1(y^B)},$$

and for the function $m(v) = v\log v$, we have

$$m(E(V)) = \frac{1}{f_1(y^A) + f_1(y^B)} K_1,$$

while working through the terms gives

$$E(m(V)) = \frac{1}{f_1(y^A) + f_1(y^B)} K_2.$$

The function m is convex, and by Jensen's inequality $E(m(v)) \geq m(E(V))$, implying $K_1 \geq K_2$. Successive application to all the KL distances in $I(f)$ means that information after transformation can be lost, but not gained. However, equality will hold for transformations for which $f_0(y^A) = f_0(y^B)$, $f_1(y^A) = f_1(y^B)$, $f_2(y^A) = f_2(y^B)$ for all y^A, y^B such that $r(y^A) = r(y^B)$.

Appendix 4: Relative Efficiency of Regression Estimates

Using the full D^2 (or D) distributions, we have for a single sibpair

$$ELOD = \frac{1}{4}E_{f_0}\log(f_0/\bar{f}\,) + \frac{1}{2}E_{f_1}\log(f_1/\bar{f}\,) + \frac{1}{4}E_{f_2}\log(f_2/\bar{f}\,),$$

where $\bar{f}$ is the population mixture of the densities, and is equal to f_1 under the bivariate normal model when $\rho_1 = (\rho_0 + \rho_2)/2$ (throughout this paper we use natural logarithms, so the result must be multiplied by $\log_e(10)$ to be compared with a LOD score). For the regression approach, we can estimate the slope by $\hat{\beta} = (1/2)(\hat{\mu}_{D_2^2} - \hat{\mu}_{D_0^2})$, where $\hat{\mu}_{D_j^2}$ is the sample mean of D^2 for sibpairs sharing j alleles IBD at the QTL. This will be very near the least squares slope estimate, as the IBD = 1 individuals primarily affect the intercept estimate. We have

$$E(\hat{\beta}) = 2\xi^2(1 - \rho_2 - (1 - \rho_0)) = 2\xi^2(\rho_0 - \rho_2)$$

and noting that D^2 is proportional to a χ_1^2 density,

$$\begin{aligned} \mathrm{var}(\hat{\beta}) &= (1/4)\{\mathrm{var}(\hat{\mu}_{D_2^2}) + \mathrm{var}(\hat{\mu}_{D_0^2})\} \\ &= (1/4)\{2(2\xi^2(1-\rho_2)) + 2(2\xi^2(1-\rho_0))^2\}/(n/4) \\ &= (8/n)\xi^4\{(1-\rho_2)^2 + (1-\rho_0)^2\}. \end{aligned}$$

For large sample sizes, and computing $t = \hat{\beta}/\sqrt{\mathrm{var}(\hat{\beta})}$, we have $t_{H0} \dot{\sim} N(0, 1)$ and $t_{HA} \dot{\sim} N(\beta/\sqrt{\mathrm{var}(\hat{\beta})}, 1)$. For two normal densities with respective means μ_A, μ_B and common variance 1, the KL distance is $(1/2)(\mu_A - \mu_B)^2$, so using the t statistic directly in a LOD calculation, we have

$$ELOD = E_{t_{HA}} \log\left(\frac{t_{HA}}{t_{H0}}\right) = (\beta^2/\mathrm{var}(\hat{\beta}))/2$$

We divide by n to obtain the contribution of a single sibpair

$$\frac{1}{16}\frac{(\rho_0 - \rho_2)^2}{(1-\rho_0)^2 + (1-\rho_2)^2}.$$

The above argument ignores some subtleties in the normal approximations, which require that the overall effect size β not be too large compared to $1/\sqrt{n}$, but suffices to develop intuition for the comparison.

Large-Sample Efficiency of the Sample Mean

For IBD(x^*) = j, we have $D \sim N(0, \xi^2(1-\rho_j))$ so $D^2 \sim c\chi_1^2$, where $c = \mathrm{var}(D)\,|\,IBD(x^*) = j) = E(D^2\,|\,IBD(x^*) = j)$. From the density of a χ_1^2 random variable and change of variable formula for $W = D^2$, we obtain the density $f_j(w) = K\, w^{1/2}\, c^{1/2}\, e^{w/2c}$, where K is a constant. For n_j independent observations with IBD(x^*) = j, solving for the maximum likelihood estimate gives the sample mean

$$\hat{c} = \frac{\sum_{i=1}^{n_j} w_i}{n_j}.$$

This is a regular likelihood setting, and so the maximum likelihood estimate is asymptotically efficient.

Appendix 5

It is possible for $I_y(f)$ to be slightly negative, which appears to be a combined consequence of a small selection region and an earlier assumption (Appendix 1) that the expected null likelihood changes little in a small neighborhood of the true gene location. However, this possibility does not appear to very important in interpreting the information measure, and so we place a lower bound of zero on the information.

References

Chang IS, Chen MY, Hsiao CF: A unified multipoint linkage analysis of qualitative and quantitative traits for sib-pairs. Stat Sinica 2002;12:297–309.

Cox DR, Hinkley DV: Theoretical Statistics. New York, Chapman and Hall, 1974, pp 1–511.

Drigalenko E: How sib pairs reveal linkage. Am J Hum Genet 1998;63:1242–1245.

Dudoit S, Speed TP: A score test for the linkage analysis of qualitative and quatitaive traits based on identity by descent data from sib-pairs. Biostatistics 2000;1:1–26.

Elston RC, Buxbaum S, Jacobs KB, Olson JM: Haseman and Elston revisited. Genet Epidemiol 2000;19:1–17.

Feingold E: Regression-based quantitative-trait-locus mapping in the 21st century. Am J Hum Genet 2002;71:217–222.

Forrest WF: Weighting improves the 'new Haseman-Elston' method. Hum Hered 2001;52:47–54.

Forrest WF, Feingold E: Composite statistics for QTL mapping with moderately discordant sibling pairs. Am J Hum Genet 2000;66:1642–1660.

Fulker DW, Cherny SS: An improved multipoint sib-pair analysis of quantitative traits. Behav Genet 1996;26:527–532.

Gaines RE, Elston RC: On the probability that a twin pair is monozygotic. Am J Hum Genet 1969;21:457–465.

Haseman JK, Elston RC: The investigation of linkage between a quantitative trait and a marker locus. Behav Genet 1972;2:3–19.

Kong A, Wright F: Asymptotic theory for gene mapping. Proc Natl Acad Sci USA 1994;91:9705–9709.

Kruglyak L, Lander ES: High-resolution genetic mapping of complex traits. Am J Hum Genet 1995;56:1212–1223.

Liang KY, Huang CY, Beaty TH: A unified sampling approach for multipoint analysis of qualitative and quantitative traits in sib pairs. Am J Hum Genet 2000;66:1631–1641.

McCulloch CE, Searle SR: Generalized, Linear and Mixed Models. New York, Wiley, 2001, pp 1–325.

Palmer LJ, Jacobs KB, Elston RC: Haseman and Elston revisited: the effects of ascertainment and residual familial correlations on power to detect linkage.Genet Epidemiol 2000 19:456–460.

Pitman EJG: A note on normal correlation. Biometrika 1939;31:9–12.

Risch N, Zhang H: Extreme discordant sib pairs for mapping quantitative trait loci in humans. Science 1995;268:1584–1589.

Sen S: Confidence intervals for gene location: The effect of model misspecification and smoothing. PhD Thesis, University of Chicago Department of Statistics, 1998.

Sham PC, Purcell S, Cherny SS, Abecasis GR: Powerful regression-based quantitative-trait linkage analysis of general pedigrees. Am J Hum Genet 2002;71:238–253.

Wang D, Lin S, Cheng R, Gao X, Wright FA: Transformation of sib-pair values for the Haseman-Elston method. Am J Hum Genet 2001;68:1238–1249.

Wang K, Huang J: A score-statistic approach for the mapping of quantitative-trait loci with sibships of arbitrary size. Am J Hum Genet 2002;70:412–424.

Wright FA: The phenotypic difference discards sib-pair QTL linkage information. Am J Hum Genet 1997;60:740–742.

Xu X, Weiss S, Xu X, Wei LJ: A unified Haseman-Elston method for testing linkage with quantitative traits. Am J Hum Genet 2000;67:1025–1028.

Zhang H, Risch N: Mapping quantitative-trait loci in humans by use of extreme concordant sib pairs: selected sampling by parental phenotypes. Am J Hum Genet 1996;59:951–957.

Human Heredity

Original Paper

Hum Hered 2003;55:143–146
DOI: 10.1159/000072319

A Simple Nonparametric Test for Linkage with Sib-Pair Censored Event Time Observations

Eric Wei[a] L.J. Wei[b] Xin Xu[c]

[a]Department of Biostatistics and Clinical Programming, Schering-Plough Research Institute, Kenilworth, N.J.,
[b]Department of Biostatistics and [c]Program for Poulation Genetics, Harvard School of Public Health,
Boston, Mass., USA

Key Words

Age onset for disease · Haseman-Elston test · Linkage analysis · Nonparametric test · Survival analysis

Abstract

Consider the case that individual phenotype and genotype observations were collected from a large or moderate number of pedigrees. Some of the pedigrees have multi-generation nuclear families. For each nuclear family, the phenotype trait value of each sibling is the time to onset for a specific event (e.g., disease). Often, this event time may be right censored, that is, an individual is event-free at the study examination time point. In this article, we propose a purely nonparametric test for testing if the distribution of a Haseman-Elston distance measure between two siblings' event times is independent of their mean genetic sharing identical by descent at a genetic marker based on such incomplete observations from all the nuclear families. The new test can be implemented easily and is illustrated with a data set from the Genetic Analysis Workshop 12. The validity of the new test is examined via a simulation study.

Introduction

In their seminal article, Haseman and Elston (H-E) [1972] proposed a novel statistical method to detect quantitative trait loci by linkage to a marker with sib-pair observations. Specifically, for two siblings in a family with trait values T_1 and T_2, let $D = (T_1 - T_2)^2$, which quantifies the distance or similarity between these two phenotype trait values. The H-E used a simple linear regression to model the relationship between the expected value, $E(D)$, of D and π, the mean genetic sharing identical by descent (IBD) at the marker, namely,

$$E(D) = \beta_0 - \beta_1 \pi, \qquad (1)$$

where β's are unknown constants. A positive β_1 indicates a possible linkage between the trait and genetic markers.

To test $\beta_1 = 0$, H-E utilized the standard least squares principle to construct a t test. If in some study families, there are multiple siblings, those squared differences may not be independent of each other. For this case, the least squares estimate for β_1 is still consistent and its standard error can be estimated well by the so-called sandwich estimate via the generalized estimating equations technique [Liang and Zeger 1986; Xu et al., 2000].

L.J. Wei
Department of Biostatistics
Harvard School of Public Health
665 Huntington Avenue, Boston, MA 02115 (USA)
E-Mail wei@hsph.harvard.edu

Instead of using D to measure the distance between the two trait values, one may use the corresponding mean-corrected product as the response variable for the above linear regression model (1) for testing linkage [Fulker and Cherny, 1996; Amos et al., 1997; Elston et al., 2000]. The resulting test performs better, but not uniformly, than the original H-E method [Elston et al., 2000]. Recently, Xu et al. [2000] proposed a powerful linkage test using both distance measures for model (1). Note that when there are a large number of relatively small nuclear families in the study, all the aforementioned tests are distribution-free for testing linkage with sib-pair observations.

Now, suppose that the phenotype variable is the time to a certain event, for example, the onset of a specific disease. Oftentimes the event time may be censored. For instance, in the Collaborative Study on Genetics of Alcoholism [Reich et al., 1998], which investigated the genetic etiology of alcoholism, 1,214 individuals from 105 families participated in this multidisciplinary, six-site study. The study was designed to identify and map genes that increase susceptibility to alcohol dependence and related phenotypes. For each individual, observations were collected for a large number of genetic markers and age at the first diagnosis of alcoholism. If an individual member was not affected at the study examination time, the age of disease onset was censored at the time of the exam. For example, if a person was 21 years old at the examination and had no drinking problem, we do not claim that this person has no drinking problem for his lifetime, but claim that his time to the event of first having a drinking problem is at least 21 years. If we are only interested in investigating if the presence or absence of alcoholism during the lifetime is linked to the marker, a standard method of utilizing a censored observation in the literature is to assume that every individual has the same susceptibility and use an estimated 'survival' function to quantify the phenotype value (the chance to have the disease during the subject's life time) for linkage analysis. For example, Zhu et al. [1997] used the above approach and performed the H-E regression method on a set of bipolar pedigrees.

An interesting and important questions is how to use the continuous trait value (the event time) directly for linkage analysis with model (1). In the presence of censored observations, it seems rather difficult if not impossible, to generalize the H-E method to handle such incomplete data. In this paper, we propose a simple, purely nonparametric method for testing linkage with possibly censored sib-pair observations. Our method is valid for the case that multi-generation nuclear families are involved. The new proposal is illustrated and validated with a data set from Genetic Analysis Workshop 12 (GAW12).

Nonparametric Test for Linkage

Let the phenotype variable T_{kia} be the time to the event of interest for the ath sib in the ith nuclear family of the kth pedigree, $a = 1, ..., K_i$; $i = 1, ..., I_k$; $k = 1, ..., n$. Let Z_k be the corresponding vector of genotypes of all the available markers in a specific chromosome of interest for all the members in the kth pedigree. Let C_{kia} be the corresponding censoring variable. One can only observe $X_{kia} = \min(T_{kia}, C_{kia})$ and an indicator Δ_{kia}, which is 1 if T_{kia} is observed, otherwise, is 0. Furthermore, assume that the vector $\{C_{kia}\}$ is independent of the vector $\{(T_{\mathrm{kia}}, Z_k)\}$.

Suppose that we are interested in a particular marker locus. For the ath and bth sibs in the ith family of the kth pedigree, let π_{kiab} be the 'estimated' number of alleles shared IBD at this marker. Let $D_{kiab} = (T_{kia} - T_{kib})^2$. Consider a rank test statistic for testing the null hypothesis that D is independent of π:

$$Q = n^{-3/2} \sum_{k=1}^{n} \sum_{l=1}^{n} \sum_{i=1}^{I_k} \sum_{j=1}^{I_l}$$

$$\left\{ \sum_{1 \le c<d \le K_j} \sum_{1 \le a<b \le K_i} I(D_{ljcd} > D_{kiab})(\pi_{kiab} - \pi_{ljcd}) \right\}, \qquad (2)$$

where $I(\cdot)$ is the indicator function and

$$\sum_{1 \le a<b \le K_i}$$

denotes a double summation of a and b, both range from 1 to K_i. A large value of Q suggests that D and π tend to have opposite directions, indicating a linkage between the trait and marker loci. Note that under the null hypothesis, the expected value of Q is zero.

Now, consider the case the T may be right censored and we only observe Δ and X, the minimum of T and C. Let $D^*_{kiab} = (X_{kia} - X_{kib})^2$. To simplify the notations, we assume that if $a < b$, $X_{kib} \ge X_{kia}$. Note that this artificial ordering does not reflect the birth order of the sibs. Now, due to censoring, not every squared sib-pair difference D^* is informative, as in the case where both sib event times T are censored and in the case when T_{kib} was observed ($C_{kib} < T_{kib}$), but the other sib's event time T_{kia} was censored ($C_{kia} > T_{kia}$). For these cases, it is not clear whether D^*_{kiab} is larger or smaller than the true D_{kiab} between these two sibs. On the other hand, D_{kiab} is at least as large as D^*_{kiab} if T_{kia} is observed and T_{kib} is censored. For this case, D^* does give us some information regarding the dissimilarity the two T's. To include all those informative squared sib-pair differences D^*, the test statistic Q in (2) may be modified to

$$Q^* = n^{-3/2} \sum_{k=1}^{n} \sum_{l=1}^{n} \eta_{kl}, \quad (3)$$

where

$$\eta_{kl} = \sum_i \sum_j \sum_{1 \le a<b \le K_i} \sum_{1 \le c<d \le K_j} \Delta_{ljc} \Delta_{kia} \Delta_{kib} I(D^*_{ljcd} > D^*_{kiab})(\pi_{kiab} - \pi_{ljcd}).$$

Note that for a non-zero summand of η_{kl}, Δ_{kia} and Δ_{kib} have to be 1, indicating that D_{kiab} is completely observed; and $\Delta_{ljc} = 1$, indicating that D_{ljcd} is at least as large as the observable D^*_{ljcd}. Therefore, if $D^*_{ljcd} > D^*_{kiab}$, this implies that $D_{ljcd} > D_{kiab}$. Under the null hypothesis, the expected value of Q^* is 0. A large value of Q^* suggests a negative correlation between D and π. The test statistic Q^* is a special case of the so-called V-statistic [Serfling, 1980]. The null distribution of Q^* can be approximated by a normal with mean 0 and variance σ^2, where

$$\sigma^2 = n^{-3} \sum_{k=1}^{n} \sum_{l=1}^{n} \sum_{k'=1}^{n} \sum_{l'=1}^{n} \left\{ \eta_{kl}\eta_{k'l'} - \sum_{\{k,l\} \neq \{k',l'\}} \eta_{kl}\eta_{k'l'} \right\},$$

where

$$\sum_{\{k,l\} \neq \{k',l'\}}$$

is a four-fold summation with indices k, l, k', l' [see, for example, Wei and Johnson, 1985, for justification of this variance estimate]. It follows that the statistic Q^*/σ can be approximated by a normal distribution with mean 0 and variance 1.

An Example

We illustrate the new distribution-free test procedure for linkage using the data from problem 2 of GAW12 [Wigsman et al., 2001]. The observations for this problem were simulated for a common oligogenic disease with five intermediate quantitative traits Q1–5, which were generated through a random effects model with various combinations of five specific genetic loci MG1–5, two environmental factors, gender and age. A quantiative disease liability was constructed as a linear function of Q1–5, MG6, and a mitochondrial component. The time to disease onset was influenced by this quantitative disease liability and a diallelic locus, MG7, on chromosome 6 at 31.5 cM. An individual who has a high disease liability value tends to have low age at onset. With this model the MG7 accounts for 30% of the variance in age at onset.

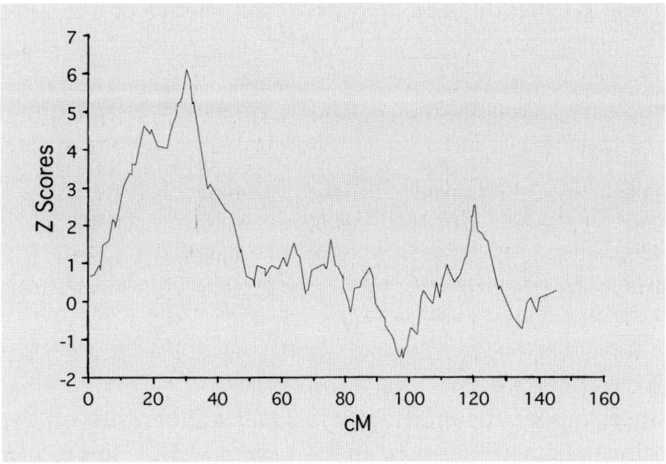

Fig. 1. Results of linkage test at 152 markers on chromosome 6 with censored age onset data from GAW12.

Both MG6 and MG7 were tightly linked on chromosome 6 near markers D06G34 and D06G35. For this simulated GAW study, there are 50 replicates of 23 large pedigrees. If a member of the pedigree is unaffected, his/her age is recorded at the study time. For illustration of our method, we selected the first ten replicates. This gives us a total of 230 pedigrees (n = 230). For each pedigree, we selected all the nuclear families and then selected all the sibs from each nuclear family. We performed the test Q^* at each of 152 markers on chromosome 6. The vector of estimated IBD sharing $\{\pi_{kiab}\}$ at 152 markers on chromosome 6 is estimated based on Z_k via the program GENIBD from S.A.G.E. package, (http://darwin.cwru.edu/pub/sage.html). Figure 1 gives a curve of the Z scores over the 152 loci. The peak of the curve is at marker D06G35, whose Z score is 6.1. Our procedure correctly identifies the marker, which is the closest one to MG7 among all the 152 markers.

This data set has also been analyzed by other novel methods, for example, proposed by Amos et al. [2001]. Specifically, Amos et al. [2001] used a Cox regression model [Cox, 1972] for survival analysis to obtain the martingale residuals and then applied the variance components procedure [Amos, 1994] to these residuals. This approach is semi-parametric and is not as robust as the present purely nonparametric procedure. On the other hand, the usage of martingale residuals may remove some confounding effects from various environmental factors (unfortunately, in contrast to the ordinary residual for the

Table 1. Empirical type-I error rates for test based on Q*

Pedigree#	Normal type-I error level			
	0.050	0.010	0.0050	0.0010
46	0.056	0.013	0.0073	0.0022
115	0.048	0.010	0.0054	0.0011
230	0.047	0.009	0.0041	0.0007
460	0.049	0.008	0.0044	0.0007

linear model, the martingale residual still depends on the adjusted covariates used in the Cox model). If the model assumptions are valid, the method by Amos et al. [2001] is expected to be more powerful than ours. It is interesting to note that our Z score curve in figure 1 has a similar profile as that given in figure 1 of Amos et al. [2001] based on the Wald test for the first 60 markers on chromosome 6.

Remarks

It is important to know whether the normal approximation to the null distribution of the test statistic Q^* is appropriate for practical sample sizes. To this end, we performed a simulation study based on the above data set. Specifically, we selected the first n pedigrees and utilized the onset times (or censored) from all the nuclear families in the GAW study. We then generated each sib pair IBD by random allele transmission from parents (whose true alleles are irrelevant), computed the test statistics Q^* and its standard error estimate σ, and recorded if Q^*/σ is greater than η_α, where η_α is the $100(1 - \alpha)$ percentage point of the standard normal. We repeated this process with 50,000 iterations. In table 1, we report the empirical type I error rate for the test based on Q^* for various combination of n and α. Note that except for the case with n = 46, the empirical levels are quite close to their nominal counterparts.

It is important to note that the validity of the normal approximation to the null distribution of the test statistic Q^* depends on the assumption that there is no ascertainment bias. If the study subjects were selected based on their phenotype trait values, but not on the genotypes, one may obtain the null distribution of the test statistic Q^* by permuting all the π's, but fixing the observed onset times (or censored) within each nuclear family. For each permutation, one obtains a realized value of Q^*. A histogram based on a large number of realized Q^* can be constructed, and the p value of the observed Q^* can be obtained accordingly. On the other hand, if there is no ascertainment bias, the test with the null distribution of Q^* based on the normal approximation described in the previous section is expected to be more powerful than the permutation test.

The new proposal can be modified easily to include other types of family relatives for testing linkage by choosing the mean genetic sharing π appropriately in Q^*.

Acknowledgements

The authors would like to thank referees for insightful and helpful comments on the paper. The test statistic was proposed in the unpublished senior-year thesis by Eric Wei at Bates College, who would like to thank his thesis advisor, Professor Melinda Harder, for her guidance during the course of the thesis research.

References

Amos CI: Robust variance-components approach for assessing genetic linkage in pedigrees. Am J Hum Genet 1994;54:535–543.
Amos CI, Krushkal J, et al: Comparison of model-free linkage mapping strategies for the study of a complex trait. Genet Epidemiol 1997;14:743–748.
Amos CI, Shete S, Gu X: Variance components analysis for genetic linkage of time to onset for disease. Genet Epidemiol 2001;21(suppl 1):S768–S773.
Cox DR: Regression models and life tables (with discussions). J R Stat Soc (B) 1972;34:187–220.
Elston RC, Buxbaum S, et al: Haseman and Elston revisited. Genet Epidemiol 2000;19:1–17.
Fulker DW, Cherny SS: An improved multipoint sib-pair analysis of quantitative traits. Behav Genet 1996;26:527–532.
Haseman JK, Elston RC: The investigation of linkage between a quantitative trait and a marker locus. Behav Genet 1972;2:3–19.
Liang KY, Zeger S: Longitudinal data analysis using generalized linear models. Biometrika 1986;73:13–22.
Reich T, Edenberg H, Goate A, et al: Genome-wide search for genes affecting the risk for alcohol dependence. Am J Med Genet 1998;81:207–215.
Serfling RJ: Approximation Theorems of Mathematical Statistics. Johns Hopkins Press, 1980.
Wei LJ, Johnson WE: Combining dependent tests with incomplete repeated measurements. Biometrika 1985;72:359–364.
Wijsman EM, Almasy L, Amos CI, Borecki I, Falk CT, King TM, Martinez MM, Meyers D, Neuman R, Olson JM, Rich S, Spence MA, Thomas DC, Vieland VJ, Wittes JS, MacCluer JW: Applications to asthma and simulated data. Genet Epidemol 2001;21(suppl 1):S1–S853.
Xu X, Weiss S, Xu X, Wei LJ: A unified Haseman-Elston method for testing linkage with quantitative traits. Am J Hum Genet 2000;67:1025–1028.
Zhu X, Olson JM, Schnell AH, Elston RC: Genetic analysis workshop 10: Model-free age-of-onset methods applied to the linkage of bipolar disorder. Genet Epidemiol 1997;14:711–716.

Original Paper

Human Heredity

Hum Hered 2003;55:147–152
DOI: 10.1159/000072320

A Method for Evaluating the Results of Bayesian Model Selection: Application to Linkage Analyses of Attributes Determined by Two or More Genes

Young Ju Suh[a,b] Kenny Q. Ye[c] Nancy R. Mendell[c]

[a]Department of Statistics, Seoul National University, and [b]Clinical Research Institute, Seoul National University Hospital, Seoul, Korea; [c]Department of Applied Mathematics and Statistics, State University of New York at Stony Brook, Stony Brook, N.Y., USA

Key Words

Linkage · Quantitative trait · Stochastic search variable selection · Intrinsic Bayes factor

Abstract

Objectives: We apply and evaluate the intrinsic Bayes factor (IBF) of Berger and Pericchi [J Am Stat Assoc 1996;91:109–122; Bayesian Statistics, Oxford University Press, vol 5, 1996] to linkage analyses done using the stochastic search variable selection (SSVS) method of George and McCulloch [J Am Stat Assoc 1993;88:881–889] as proposed by Suh et al. [Genet Epidemiol 2001;21(suppl 1):S706–S711]. ***Methods:*** We consider 20 simulations of linkage data obtained under two different generating models. The SSVS is applied to a multiple regression extension [Genet Epidemiol 2001;21(suppl 1): S706–S711] of the Haseman-Elston [Behav Genet 1972;2:3–19; Genet Epidemiol 2000;19:1–17] methods. Four prior distributions are considered. We apply the IBF criterion to those samples where different prior distributions result in different top models. ***Results:*** In those samples where three different models were obtained using the four priors, application of the IBFs eliminated one of the two wrong models in 4 out of 5 situations. Further elimination using the IBF criterion for situations with two different subsets did not serve as well. ***Conclusions:*** When different priors result in three or more different subsets of markers, one can use the IBF to get this number down to two for consideration. When two subsets result we recommend that both be considered.

Introduction

Familial traits are generally governed by genetic and non-genetic factors. In order to determine whether there is a major gene for a trait linked to some well-defined marker, linkage analysis is performed on the joint distribution of the trait and the marker alleles in family members. Alternatively association studies are done in which the distribution of alleles at polymorphic markers is compared in cases and normal controls (often unaffected members of the case family). In the case of quantitative traits, linkage analysis involves the investigation of the correlation between trait identity and identity at major genetic markers. A very popular genetic-model-free meth-

KARGER

Fax +41 61 306 12 34
E-Mail karger@karger.ch
www.karger.com

Accessible online at:
www.karger.com/hhe

Dr. Nancy R. Mendell
Department of Applied Mathematics and Statistics
State University of New York at Stony Brook
Stony Brook, NY 11794-3600 (USA)
Tel. +1 631 632 8373, Fax +1 631 632 8490, E-Mail nmendell@notes.cc.sunysb.edu

od for analysis of quantitative traits was proposed by Haseman and Elston [1]. This method involves regression of the measure's squared difference in pairs of siblings on the number of alleles identical by descent (i.b.d.) at a given marker. Elston et al. [2], proposed instead of regression the measure's trait cross product in pairs of siblings on the number of alleles i.b.d. at a given marker so as to assure non-zero covariance between observations obtained when two sib pairs in the sample involve a common individual. (This occurs whenever a sibship in the sample consists of three or more individuals.). Since this time further improvements of the Haseman-Elston method have been proposed (Forrest and Feingold [3], Xu et al. [4], Wang et al. [5]). All of these methods can be seen as regression based methods (RBM) for linkage analysis. Their application has involved only univariate analysis of one marker locus at a time. Given that it is likely that these traits are due to more than one gene (or could be associated with more than one allele at a given locus) these methods can and should be extended to multiple regression methods (or, in the case of a qualitative trait, multiple logistic regression methods) so that associations at several loci can be considered simultaneously. Also, epistasis of major genes might be considered through addition of cross-product terms.

However, the problem of finding the optimal subset among all the possible combinations of candidate predictors involves considering a large number of marker combinations. Upon adding all possible two-way interactions (in order to allow for epistasis), the number of possible subsets increases to an even greater value. To overcome this difficulty in the analysis of a large set of predictor variables, a stochastic search variable selection (SSVS) approach based on Bayesian regression was proposed by George and McCulloch [6]. Suh et al. [7] extended the method of George and McCulloch [6] to the linkage analysis problem as a method for evaluating data on several marker loci simultaneously. The SSVS was reasonably successful in identifying the correct subset of linked markers for both of these traits in the study by Suh et al. [7]. However it was noted at the time that different priors resulted in somewhat different subsets of markers being identified. Additionally it was noted that the posterior probabilities of all models were quite low.

George and McCullough [6] indicated that the choice of prior might matter and suggested that users should experiment by varying the values of the parameters of the prior distribution, keeping the problem being addressed in mind. They suggested that one needed agreement in conclusions with several different priors in order to have confidence in one's findings. Beattie et al. [8] combined SSVS with the intrinsic Bayes factor (*IBF*) proposed by Berger and Pericchi [9, 10], to develop a two-stage approach for analyzing supersaturated designs. Here we apply the same approach to the linkage analysis with a RBM. Our goal in this study is to evaluate the effectiveness of the *IBF* approach when applied to several models obtained from SSVS with different priors. Our criterion here is the success of this combined approach in correctly identifying linked markers for traits that are due to more than one major gene.

Methods

The Application of Stochastic Search Variable Selection to Regression-Based Linkage Analysis

The method proposed by George and McCulloch [6] applies to regression of a dependent variable Y on p possible predictors X_1, X_2, ..., X_p. With this method each model is represented by a binary vector $\gamma = (\gamma_1, ..., \gamma_p)$. If a variable X_i is in the model, $\gamma_i = 1$; otherwise, $\gamma_i = 0$. The slope of X_i has prior distribution $N(0, c\tau^2)$ when $\gamma_i = 1$, and prior distribution $N(0, \tau^2)$ when $\gamma_i = 0$. Here, c is greater than 1. The result of the analysis is the posterior distribution of all vectors γ (of dimension p). One applies the method and then selects the model with highest of the estimated posterior probabilities obtained for all of the candidate models. That is, the optimal model is identified by the value of the γ with the highest posterior probability, referred to as, $\gamma^+ = (\gamma_1^+, ..., \gamma_p^+)$. George and McCulloch [6] proposed the following approach: (1) Perform Bayesian regression analysis using the priors described above. (2) Determine the posterior distribution of γ. (3) Find the value of γ having the highest posterior probability, $\gamma^+ = (\gamma_1^+, \gamma_2^+, ..., \gamma_p^+)$. (4) The optimal subset for predicting Y consists of those X_i for which $\gamma_i^+ = 1$. Suh et al. [7] modified the method of George and McCulloch [6] through applying a prior of $N(h\tau^2, c\tau^2)$ for the case where $\gamma = 1$. As suggested by George and McCulloch [6], Suh et al. [7] used the semiautomatic settings $(\sigma_{\beta i}/\tau_i, c_i)$ = (1, 5), (1, 10), (10, 100) and (10, 500) where $\sigma_{\beta i}$ denotes the estimated standard error of β_i based on a univariate model.

Further details on how to apply the George and McCulloch SSVS [6] method to regression based linkage methods in general and to the Elston et al. [2] method in particular are given in Suh et al. [7].

Intrinsic Bayes Factor Analysis

The Bayes factor is a Bayesian statistic for comparing two models. One advantage of the Bayes factor is that it can be used to compare non-nested models (whereas in the frequentist hypothesis testing approach one model has to be nested in the other). However, the successful application of the Bayes factor requires a proper prior to be assigned to the parameter space, since non-informative priors are equivalent to multiplication by a constant, giving meaningless values to the Bayes factors. The main purpose of computing the *IBF* [9, 10] is to overcome this deficiency and obtain automatically a Bayes factor which corresponds to what one would obtain using standard non-informative priors.

In this study, the *IBF* approach is applied to the two competing linear models, say M_1 and M_2. With the *IBF*, a constant prior can be used, and there is no need to choose a proper prior for the parameters. The IBF of M_2 against M_1 is as follows,

$$IBF_{21}(l) = BF_{21}(Y(-l)|Y(l))$$

$$= \frac{\int f(Y(-l)\,|\,\theta_2, Y(l), M_2)\pi^N(\theta_2\,|\,Y(l), M_2)d\theta_2}{\int f(Y(-l)\,|\,\theta_1, Y(l), M_1)\pi^N(\theta_1\,|\,Y(l), M_1)d\theta_1}$$

$$= \frac{\int f(Y\,|\,\theta_2, M_2)\pi^N(\theta_2\,|\,M_2)d\theta_2}{\int f(Y\,|\,\theta_1, M_1)\pi^N(\theta_1\,|\,M_1)d\theta_1} \cdot \frac{\int f(Y(l)|\,\theta_1, M_1)\pi^N(\theta_1\,|\,M_1)d\theta_1}{\int f(Y(l)|\,\theta_2, M_2)\pi^N(\theta_2\,|\,M_2)d\theta_2}$$

$$= BF_{21}^N \cdot BF_{12}^N(l).$$

Here $Y(l)$ denotes a part of data, and $Y(-l)$ indicates the remaining data in the sample where $Y = (Y(l), Y(-l))$, called the 'minimal training sample' with a sample size of $m = \max\{k_1, k_2\} + 1$, where k_i denotes the number of parameters (including an intercept term) of model M_i $(i = 1, 2)$, and $\pi^N(\theta_i|M_i)$ denotes the non-informative prior for the full dataset Y under M_i $(i = 1, 2)$.

Berger and Pericchi [9, 10] suggested two types of *IBF*s, which are the 'arithmetic mean of the partial intrinsic Bayes factor' (*AIBF*) and the 'geometric mean of the partial intrinsic Bayes factor' (*GIBF*). These are obtained by averaging over all minimum training samples. Detailed discussions on the advantage and disadvantage of these two types of *IBF*s were also given in [9, 10]. Among choices of non-informative priors, we focused on the reference prior, as suggested in Berger and Pericchi [9, 10]. The corresponding *AIBF* obtained by using the reference prior is then

$$AIBF_{21} = \frac{|X_1^t X_1|^{1/2}}{|X_2^t X_2|^{1/2}} \cdot \frac{R_1^{(n-k_1)/2}}{R_2^{(n-k_2)/2}} \cdot \frac{C}{L}\sum_{l=1}^{L} \frac{|X_2^t(l)X_2(l)|^{1/2}}{|X_1^t(l)X_1(l)|^{1/2}} \cdot \frac{R_2(l))^{1/2}}{(R_1(l))^{(k_2-k_1+1)/2}} \qquad (1)$$

where X_i is an $n \times k_i$ matrix of rank k_i $(< n)$, R_i is the residual sum of squares for the model M_i $(i = 1, 2)$, $C = \{\Gamma((n-k_2)/2)\Gamma((k+1)/2)\}/\{\Gamma((n-k_1)/2)\Gamma(1/2)\}$, and L is the total number of minimal training samples. The *GIBFs* can be similarly obtained. Our decision-making using the *IBF* is based on the recommendation given for decision-making by Jeffreys [11] and Kass and Raftery [12]. That is, if $IBF > 3$, select model 1, if $IBF < 1/3$, select model 2, otherwise consider both models.

Beattie et al. [8] noted that the two Bayesian model selection procedures, SSVS and *IBF*, compliment each other. They proposed applying *IBF* to the top models selected by SSVS, and applied this two-stage approach to the analysis of designed experiments. In this study, we apply the same approach to genetic linkage analysis.

Data Sets Used for the Evaluation of the Effectiveness of the *IBF*

The data was provided by the 12th Genetic Analysis Workshop (GAW12, Problem 2: simulated dataset) [13]. It consisted originally of 50 replicates with each replicate containing 23 extended pedigrees consisting of a total of 1,497 individuals. In all 50 replicates the distribution of pedigree structures is identical. Each pedigree has at least three children. For each individual, data is provided on 22 autosomes, with 58 to 220 markers per chromosome, resulting in data on a total of 2,855 markers per individual. In this study we do linkage analyses with the objective of identifying those markers which are linked to the major genes for two quantitative traits which we refer to as Q1 and Q2. Both of these traits are determined by more than one gene and other non-genetic factors. In reality, one would first adjust for known associated non-genetic factors before proceeding with a genetic analysis. In the case of these simulated data it was 'known' that Q1 depends on age and gender and Q2 on age, gender, and an environmental factor. Thus the results reported here and at the workshop were based on the analysis of age-gender-adjusted values of Q1 which we refer to as Q1*, and age-, gender- and environment-adjusted values of Q2, as Q2*. Details on how these adjustments were determined are given in Suh et al. [7]. For our study, we combined GAW12 simulated samples 1–5, 6–10, ..., 46–50 into 10 'super samples', each with n = 3,855 sib-pairs (as we did in Suh et al. [7]). The model selection procedures are conducted on these 10 super samples.

Table 1. Q1* and 12 markers[a]: Comparison of the results of stochastic search variable selection under different prior distributions (correct model is I_1, I_2)

Sample[b]	(1, 5)[c]	(1, 10)	(10, 100)	(10, 500)	Number of models
1	I_1, I_2, U_1	I_1	I_1, I_2^e	I_1	3
2	I_2	I_2	I_2	I_2	1
3	I_1, I_2	I_1, I_2	I_1, I_2	I_2	2
4	I_1, I_2	I_1, I_2	I_1, I_2	I_1, I_2	1
5	I_1, I_2	I_1, I_2	I_1, I_2	I_1	2
6	I_1, I_2	I_2	I_1, I_2	I_2	2
7	I_1, I_2	I_1, I_2	I_1, I_2	I_1, I_2	1
8	I_1, I_2	I_1, I_2	I_1, I_2, U_6	I_1	3
9	I_1, I_2	I_2	I_1, I_2, U_9	I_2	3
10	I_1, I_2	I_1, I_2	I_1, I_2, U_6	I_1, I_2	2

[a] 10 unlinked markers (U_i) plus two linked markers (I_1, I_2) are considered.
[b] Simulations with 3,855 sib pairs in each sample.
[c] The values on legend indicate ($\sigma_{\beta i}/\tau_i$, c_i) in the prior of β_i.

In the analysis by Suh et al. [7] we were interested in considering how well the SSVS worked in 'a best case scenario'. That is we wanted to see how well this method performed when we considered as predictors the information on the number of alleles i.b.d. at the 2 markers closest to MG1 (I_1) and MG2 (I_2) in the case of Q1. In the case of Q2, which was determined by additive effects at MG1 and MG2 and epistasis at MG2 and MG3, the candidate predictors considered are the number of alleles i.b.d. at the markers closest to MG1 (I_1), MG2 (I_2) and MG3 (I_3) and the cross product of I_2 and I_3 (I_{2*3}) as a candidate predictor. We considered only 10 of the markers which are not linked to MG1, MG2 and MG3. These markers are unlinked to each other. This was done to evaluate the efficacy of this method in the case where we have no collinearity. Thus the remaining candidate predictors for both Q1 and Q2 are U_1, U_2, ..., U_{10} where U_j denotes the identity by descent at the jth unlinked marker considered.

The Problem

Tables 1 and 2 show the top models obtained using the four different semiautomatic priors suggested by George and McCulloch

Table 2. Q2* and 14 markers[a]: Comparison of the results of stochastic search variable selection under different prior distributions (correct model is I_1, I_2, I_3, I_{2*3})

Sample[b]	(1, 5)[c]	(1, 10)	(10, 100)	(10, 500)	Number of models
1	I_2, I_3	I_2, I_3	I_2, I_3	I_2, I_3	1
2	I_2, I_3	I_2, I_3	I_2, I_3	I_2, I_3	1
3	I_1, I_2	I_2	I_1, I_2, I_3, I_{2*3} [e]	I_2	3
4	I_2, I_3	I_2, I_3	I_2, I_3	I_2, I_3	1
5	I_2, I_3, I_{2*3}, U_3	I_3	I_2, I_3, I_{2*3}, U_3	I_3	2
6	I_2, U_2	I_2, U_2	I_2, U_2	I_2, U_2	1
7	I_1, I_2, I_3, I_{2*3}	I_1, I_2, I_3, I_{2*3}	I_1, I_2, I_3, I_{2*3}	I_1, I_2, I_3, I_{2*3}	1
8	I_2, I_3, I_{2*3}	I_2, I_3, I_{2*3}	I_1, I_2, I_3, I_{2*3}	I_2, I_3	3
9	I_2	I_2	I_2	I_2	1
10	I_2, I_3	I_2, I_3	I_2, I_3	I_2, I_3	1

a 10 unlinked markers (U_i) plus 3 linked markers and epistastis between two markers (I_1, I_2, I_3 and I_{2*3}) are considered.
b Simulations with 3,855 sib pairs in each sample.
c The values on legend indicate ($\sigma_{\beta i}/\tau_i$, c_i) in the prior of β_i.

[6] for ($\sigma_{\beta i}/\tau_i$, c_i). The important thing to note is that in the case of Q1* we have 8/10 situations where different priors result in different top models and in the case of Q2* we have 3/10 situations where different priors result in different top models. In the case of Q1*, the prior with ($\sigma_{\beta i}/\tau_i$, c_i) set at (1, 5) appeared to be the best one with respect to the power to detect the correct subset (I_1, I_2) and having the lowest rate of false positives (choice of marker genes unlinked to MG1 and MG2). For Q2*, the prior set at (10, 100) most frequently resulted in the correct model (I_1, I_2, I_3, I_{2*3}), and (10, 500) slightly reduced the observed number of false positive results. However, under the choice of ($\sigma_{\beta i}/\tau_i$, c_i) at (1, 5), the largest posterior probability ('peak' signal, the one of the chosen optimal model γ^+) among all models was small with average values of 0.045 for Q1* and 0.022 for Q2* in these 10 samples, which indicates very 'flat' posterior distributions. When the prior at (10, 500) was used, the power to identify linked markers was low for both Q1* and Q2*.

The purpose of this study is to apply the *IBF* to these 10 simulations where we obtained different top models using different priors and evaluate how well it works.

Results

The *IBF* analysis was applied in an attempt to resolve the fact that different prior distributions resulted in different top models in 10 out of 20 datasets. Pairwise *IBF* analyses were applied to the top models chosen from each of the four priors. Because in our samples no more than three different top models were obtained upon using four different priors, at most three pairwise analyses needed to be done for a given simulated sample. In the case where there are only two different top models under different priors, one *IBF* analysis was done. The *AIBF* and the *GIBF* were computed using the reference prior. Following the suggestions given by Pericchi [pers. commun.], we considered the prior under the minimal training samples of $L = n \times m_k$ in equation (1), where n = 3,855 observations and m_k is the minimal training sample size for the kth super sample (k = 1, ..., 10).

As shown in table 3, in the cases where we obtained three different top models with four different priors, we were able to eliminate at least one of the models via application of the *IBF*. For example (in super sample 1 for Q1*), we have the IBF_{21} = 50.6, IBF_{31} = 38.8 and IBF_{32} = 1.6 for the three different models obtained with the three different priors, where IBF_{ij} indicates the *IBF*s of model i against model j. It is clear from the values of the IBF_{21} and the IBF_{31} we can consider Model 1 as much weaker than these two other models. Therefore, Model 1 is discarded in this case. Meanwhile the IBF_{32} is close to 1.0 which indicates that models 2 and 3 are comparable. In two of these cases, each for Q1* and Q2*, the eliminated model was missing a predictor variable which should be included while the other two contained this predictor variable.

After eliminating one of the three models, we applied *IBF* analyses of two remaining competing models, together with the cases in which two different top models were obtained with four different priors. Upon considering the 10 cases where we had two models, our results were unambiguous (*IBF* >3 or $<1/3$) in 6/10 these cases and ambiguous results in the 4/10 cases (using the *GIBF* with the reference prior). For the *AIBF* these values were 2/10 and 8/10 respectively. However in 4/6 of the definitive situations the result of the *GIBF* analysis was the choice of

Table 3. Results from each of the 3 pairwise intrinsic bayes factor analyses[a] considering the top models obtained upon using 4 different priors[b] (correct model for Q1* is I_1, I_2; correct model for Q2* is I_1, I_2, I_{2*3})

Sample[c]	*AIBF* (A) and *GIBF* (G)			Model 1[d]	Model 2	Model 3
Q1*	IBF_{21}[e]	IBF_{31}	$\boldsymbol{IBF_{32}}$	√		
1	(A) 50.61	38.83	**1.6**	I_1, I_2, U_1	I_1, I_2	I_1
	(G) 112.28	265.14	**3.6**			
Q1*	$\boldsymbol{IBF_{12}}$	IBF_{13}	IBF_{32}			√
8	(A) **5.75**	89,008	26,438	I_1, I_2	I_1, I_2, U_6	I_1
	(G) **12.77**	40,994.7	4,878.26			
Q1*	IBF_{12}	$\boldsymbol{IBF_{31}}$	IBF_{32}		√	
9	(A) 2.68	**2.44**	3.45	I_1, I_2	I_1, I_2, U_9	I_2
	(G) 5.97	**5.46**	21.27			
Q2*	IBF_{21}	IBF_{31}	$\boldsymbol{IBF_{23}}$	√		
3	(A) 7.24	40,484.9	**2.62**	I_1, I_2	I_1, I_2, I_3, I_{2*3}	I_2
	(G) 17.68	9.5	**0.53**			
Q2*	IBF_{12}	$\boldsymbol{IBF_{31}}$	IBF_{32}		√	
8	(A) 4.72	**1.57**	3.85	I_2, I_3, I_{2*3}	I_1, I_2, I_3, I_{2*3}	I_2, I_3
	(G) 11.21	**4**	29.32			

[1] The *AIBF* and *GIBF* analyses are considered by using the reference prior.
[b] In the case where there are 3 different top models under different priors of $(\sigma_{\beta i}/\tau_i, c_i) = (1, 5), (1, 10), (10, 100)$ and $(10, 500)$.
[c] Samples for which we obtain 3 different top models using different prior distributions.
[d] 3 different top models selected from the method of George and McCulloch using 4 different priors.
[e] IBF_{ij}: The *IBF* of model l *i* against model *j*. Written in bold fonts indicates the selected *IBF* with the smallest value of those *IBF*s of each sample.
√ An eliminated model, corresponding to the *IBF* selected in each sample (see methods).

the wrong model. For this reason we conclude that if one obtains only two different top models with different priors and/ or one has narrowed it down through application of the *AIBF* or *GIBF* from four or three to two top models, one should investigate each of these two top models.

We note also that in our situation, application of the *IBF* to the case where there are three or more top models is likely to eliminate an incorrect one. We do need to allow however, for the fact that we only have 10 situations where different priors resulted in different models (as evaluated in tables 1–3) and only 5 where different priors resulted in three different models. Thus we hardly have sufficient evidence as to the information provided by application of the *IBF* to the genome-wide search problem where we have 200 possible loci to consider. However we find these results promising. We do have evidence in fact that the application of this method may clarify the findings (in the case of three different top models) and it will not compound the error of the analysis.

Discussion

The method of George and McCulloch [6] focuses on selecting a promising model identified by the highest posterior probability out of all candidate models by using the SSVS procedure. The method has an advantage in terms of reducing the computational burden, compared to the classical 'frequentist' best subset selection technique which requires considering of all 2^p possible subsets. In applying this method to our datasets, four different priors were implemented with the $p = 12$ independent predictors considered for Q1* and $p = 14$ independent predictors considered for Q2* in 10 simulated samples. This method had good power for choosing the correct subset of loci most of the time [7]. However, results using this method were rather sensitive to the choice of the parameters in the prior distribution. In fact, different priors gave slightly different top models in 7/10 cases for Q1*. Additionally, the top two models selected for each prior had posterior probabilities which are very close to each other. George and McCulloch [6] proposed applying other standard model selection approaches to get a unique promising model. In this regard, as a possible modification to evaluation of George and McCulloch's method [6], we considered the *IBF* technique [9, 10]. The *IBF* approach was applied pairwise to the top models obtained using each pair wherever the four priors gave different top models. Upon doing this (*IBF*) we obtained unambiguous results (i.e. the *IBF* resulted in a decision in favor of two models over a third in 5/5 cases where we had three top models). We correctly eliminated the incorrect model in 4/5 of these cases (i.e. sample 8 for Q2* shown in table 3). Thus we conclude, based on this *very small simulation*, that the *IBF* might be used for excluding subsets in the case where we have three or more top models with different priors. However once one has two top models with different priors, one should consider both.

Nevertheless, this can be seen only as a preliminary finding with respect to the utility of applying *IBF* to this problem. We have considered the case where only 10 unlinked loci are considered along with the correct marker alleles. The utility of the SSVS method in a more realistic situation where we consider over 200 markers and hence have collinear candidate predictors needs to be addressed as well as how well the *IBF* method works when one obtains competing models using different priors in this case. However, one first needs to see how well these methods perform in optimal conditions before investing in the more expensive analysis and that is what we have done here. On the other hand we expect that these combined techniques might be even more useful when applied to linkage data in which the major genes account for a greater percentage of variance. To the extent that some other regression-based linkage approach is more sensitive than the one we used, these results might be slightly better had we used a different regression-based approach. However that is not the point of this paper. The point here is to propose and evaluate a method to address an inevitable problem inherent in using SSVS, namely that different priors result in different top models. The methods proposed here apply to all regression-based linkage methods.

For future studies we propose to do a larger simulation and examine the behavior of the two-stage analysis combining SSVS [6] and the *IBF* approach [9, 10] under stronger models, i.e. models in which the predictors account for a large percentage of the variance. We believe that if there is a stronger relation between the number of alleles i.b.d. and degree of trait sharing, the method might perform better than it did in this case. When interaction effects (epistasis) are considered, we will consider the hierarchical model priors in SSVS as proposed by Chipman [14]. Additionally, Beattie et al. [8] showed that the SSVS/*IBF* approach works well for supersaturated designs. Therefore, we'll consider applications in which the number of marker genes is greater than the number of observations.

Acknowledgements

We thank Dr. Stephen J. Finch for his advice regarding linkage analysis, Dr. Wentian Li for his comments, Dr. Luis Pericchi for generously providing his FORTRAN programs for computing intrinsic Bayes factors, and GAW12 data providers for the permission to use the data. The first author acknowledges grant Brain Korea 21 from the Korea Research Foundation, the second author acknowledges grant DMS 0306306 from the National Science Foundation, and the third author acknowledges grant MH49487 from the National Institute of Mental Health. The National Institute of General Medical Sciences grant GM31575 supports GAW.

References

1 Haseman JK, Elston RC: The investigation of linkage between a quantitative trait and a marker locus. Behav Genet 1972;2:3–19.
2 Elston RC, Buxbaum S, Jacobs KB, Olson JM: Haseman and Elston revisited. Genet Epidemiol 2000;19:1–17.
3 Forrest WF, Feingold E: Composite statistics for QTL mapping with moderately discordant sibling pairs. Am J Hum Genet 2000;66:1642–1660.
4 Xu Xin, Weiss S, Xu Xiping, Wei LJ: A unified Haseman-Elston method for testing linkage with quantitative traits. Am J Hum Genet 2000;67:1025–1028.
5 Wang D, Lin S, Cheng R, Gao X, Wright FA: Transformation of sib-pair values for the Haseman-Elston method. Am J Hum Genet 2001;68:1238–1249.
6 George EI, McCulloch RE: Variable selection via Gibbs sampling. J Am Stat Assoc 1993;88:881–889.
7 Suh YJ, Finch SJ, Mendell NR: Application of a Bayesian method for optimal subset regression to linkage analysis of Q1 and Q2. Genet Epidemiol 2001;21(suppl 1):S706–S711.
8 Beattie SD, Fong DKH, Lin DKJ: A two-stage Bayesian model selection strategy for supersaturated designs. Technometrics 2002;44(1):55–63.
9 Berger JO, Pericchi LR: The intrinsic Bayes Factor for linear models; in Bernardo JM, Berger JO, David AP, Smith AFM (eds): Bayesian Statistics. Oxford, Oxford University Press, 1996, vol 5, pp 23–42.
10 Berger JO, Pericchi LR: The intrinsic Bayes factor for model selection and prediction. J Am Stat Assoc 1996;91:109–122.
11 Jeffreys H: Theory of Probability, ed. 3. Oxford, Oxford University Press, 1961.
12 Kass RE, Raftery AE: Bayes factors. J Am Stat Assoc 1995;90:773–795.
13 Wijsman EM, Almasy L, Amos CI, Borecki I, Falk CT, King TM, Martinez MM, Meyers D, Neuman R, Olson JM, Rich S, Spence MA, Thomas DC, Vieland VJ, Witte JS, MacCluer JW: Analysis of complex genetic traits: Applications to asthma and simulated data. Genet Epidemiol 2001;21(suppl 1):S1–S853.
14 Chipman H: Bayesian variable selection with related predictors. Can J Stat 1996;24:17–36.

Update your knowledge with these expert reviews

Vertebrate Sex Chromosomes

Guest Editor:
Nobuo Takagi

Genetics, Developmental Biology, Embryology, Molecular Biology, Zoology

During the last two decades of the 20th century, the study of chromosomes was enormously stimulated by the availability of various biological resources, valuable databases and new genetic and molecular biological techniques including PCR and fluorescent in situ hybridization. The simultaneous development of new and effective systems for targeting genes and gene transfer led to astonishing advances also in the study of vertebrate sex chromosomes. We now have considerable insight into their evolution, the dosage compensation mechanisms in mammals that involve chromosome-wide chromatin remodeling triggered by noncoding Xist RNA, and the unique properties of the human X chromosome which comprises many genes important for sex determination and reproduction.
This thematic issue of *Cytogenetic and Genome Research* presents an excellent and highly authoritative compilation of current research findings related to vertebrate sex chromosomes. The contributions – all prepared by experts in the field – reflect both the technical advances and the enormous gain in knowledge achieved in recent years. Many of these papers also provide important historical perspectives while some contain insightful personal views of the authors. This special volume therefore represents a stimulating and critical state-of-the-art source of information for investigators and students in this fascinating field.

Selected Contents

X Chromosome Inactivation
X chromosome inactivation: theme and variations: ***Migeon, B.R.***
X chromosome inactivation, differentiation, and DNA methylation revisited, with a tribute to Susumu Ohno: ***Riggs, A.D.***
ICF syndrome cells as a model system for studying X chromosome inactivation: ***Gartler, S.M.; Hansen, R.S.***
X chromosome inactivation: how human are mice?: ***Vasques, L.R.; Klöckner, M.N.; Pereira, L.V.***
Escape from X inactivation: ***Disteche, C.M.; Filippova, G.N.; Tsuchiya, K.D.***

Evolution of Sex Chromosomes
The evolution of sex chromosomes: ***Ayling, L.-J.; Griffin, D.K.***
The Azoospermia region AZFa: An evolutionary view: ***Wimmer, R.; Kirsch, S.; Weber, A.; Rappold, G.A.; Schempp, W.***
Early origins of the X and Y chromosomes: Lessons from tilapia: ***Griffin, D.K.; Harvey, S.C.; Campos-Ramos, R.; Ayling, L.-J.; Bromage, N.R.; Masabanda, J.S.; Penman, D.J.***
Chromosome evolution in fish: sex chromosome variability in *Eigenmannia virescens* (Gymnotiformes: Sternopygidae): ***de Almeida-Toledo, L.F.; Daniel-Silva, M.F.Z.; Moysés, C.B.; Fonteles, S.B.A.; Lopes, C.E.; Akama, A.; Foresti, F.***

Development and Sex Determination
Mammalian embryonic development – insights from studies on the X chromosome: ***Monk, M.***
The X chromosome and ovarian function: ***Vialard, F.; Cocquet, J.; Christin-Maitre, S.; Veitia, R.; Fellous, M.***
The role of *Zic3* in vertebrate development: ***Herman, G.E.; El-Hodiri, H.M.***

X-Linked Diseases
The X chromosome and fragile X mental retardation: ***Oostra, B.A.; Willemsen, R.***
The proportion of cells with functional X disomy is associated with the severity of mental retardation in mosaic ring X Turner syndrome females: ***Kubota, T.; Wakui, K.; Nakamura, T.; Ohashi, H.; Watanabe, Y.; Yoshino, M.; Kida, T.; Okamoto, N.; Matsumura, M.; Muroya, K.; Ogata, T.; Goto, Y.; Fukushima, Y.***

Sex Chromosome Organization
X-chromosomal localization of mammalian Y-linked genes in two XO species of the Ryukyu spiny rat: ***Arakawa, Y.; Nishida-Umehara, C.; Matsuda, Y.; Sutou, S.; Suzuki, H.***
Biallelic expression of Z-linked genes in male chickens: ***Kuroiwa, A.; Yokomine, T.; Sasaki, H.; Tsudzuki, M.; Tanaka, K.; Namikawa, T.; Matsuda, Y.***

www.karger.com/cgr

KI 03271

Cytogenetic and Genome Research
Editors: Schmid, M. (Würzburg); Klinger, H.P. (New York, NY)
ISSN 1424–8581
Hardcover Reprint: Vol. 99, No. 1–4 (2002)

Vertebrate Sex Chromosomes
Editor: Takagi, N. (Sapporo)
352 p., 171 fig., 58 in color, 54 tab., hard cover, 2003
CHF 159.– / EUR 113.50 / USD 138.25
Prices subject to change
EUR price for Germany, USD price for USA only
ISBN 3–8055–7637–4

Order Form

Please send: ____ copy/ies
Postage and handling free with prepayment

Payment:
☐ Check enclosed ☐ Please bill me

Please charge to my credit card
☐ American Express ☐ Diners ☐ Eurocard
☐ MasterCard ☐ Visa

Card No.: ____________________

Exp. date: ____________________

Fax: +41 61 306 12 34
S. Karger AG, P.O. Box, CH–4009 Basel (Switzerland)
E-Mail orders@karger.ch, **www.karger.com**

Name/Address: ____________________

Date: ____________________

Signature: ____________________

KARGER

Orders may be placed with any bookshop, subscription agency, directly with the publisher or through a Karger distributor.